SPSS® For Dummies

P9-CES-556

Sheet

Variable Definitions

Name: Short form of the variable name

Type: Numeric, comma, dot, scientific notation, date, dollar, custom currency, or string

Width: Maximum number of characters used to display the data

Decimals: Number of digits to the right of the decimal point

Label: Long form of the variable name

Values: Names assigned to specific values

Missing: Value, or values, to represent missing values

Columns: Number of spaces into which the value is displayed

Align: Right, left, or center

Measure: Scale, ordinal, or nominal

Syntax Language Statements

command [/*option=value*] *terminator*

command: Every statement begins with a command.

option: Each command has a specific set of options.

value: The value, or values, for the option.

terminator: Every statement ends with a period as a terminator.

Syntax Language Predefined Variables

Variable Name	What It Is
$CASENUM	Current case number. It is the count of cases from the beginning to the current one.
$DATE	Current date in international date format.
$JDATE	Count of the number of days since October 14, 1582 (the first day of the Gregorian calendar).
$LENGTH	Current page length.
$SYSMIS	System missing value. This prints as a period (.) or whatever is defined as the decimal point.
$TIME	Number of seconds since midnight October 14, 1582 (the first day of the Gregorian calendar).
$WIDTH	Current page width.

SPSS® For Dummies®

Syntax Language Comments

```
COMMENT This is a comment and will not be executed.
* This is a comment and will continue to be
  a comment until the terminating period.
/* This is a comment and will continue to be
  a comment until the terminating asterisk-slash */
```

Syntax Language Relational Operators

Symbol	Alpha	What It Is
=	EQ	Is equal to
<	LT	Is less than
>	GT	Is greater than
<>	NE	Is not equal to
<=	LE	Is less than or equal to
>=	GE	Is greater than or equal to

Syntax Language Logical Operators

Symbol	Alpha	Definition
&	AND	Both relational operators must be true
|	OR	Either relational operator can be true
~	NOT	Reverses the result of a relational operator

For Dummies: Bestselling Book Series for Beginners

SPSS® FOR DUMMIES®

by Arthur Griffith

Wiley Publishing, Inc.

SPSS® For Dummies®

Published by
Wiley Publishing, Inc.
111 River Street
Hoboken, NJ 07030-5774

www.wiley.com

Copyright © 2007 by Wiley Publishing, Inc., Indianapolis, Indiana

Published by Wiley Publishing, Inc., Indianapolis, Indiana

Published simultaneously in Canada

For general information on our other products and services, please contact our Customer Care Department within the U.S. at 800-762-2974, outside the U.S. at 317-572-3993, or fax 317-572-4002.

For technical support, please visit www.wiley.com/techsupport.

Wiley also publishes its books in a variety of electronic formats. Some content that appears in print may not be available in electronic books.

Library of Congress Control Number: 2006939501

ISBN: 978-0-470-11344-8

Manufactured in the United States of America

10 9 8 7 6 5 4 3

About the Author

Arthur Griffith is a computer programmer and a writer. He is the author of eleven books and the coauthor of three. His education was many years ago in a land far away, and he has a degree in Computer Science and Mathematics.

During his years as a computer programmer, he developed systems as varied as nuclear power plant construction accounting, missile guidance, remote control of cable-TV set-top boxes, and satellite communications control. All the work he did with computer programming required the use of mathematics and the ability to explain complex concepts in simple language.

He moved to Alaska in an attempt to retire, but failed. He is now developing software for Kachemak Research Development and writing books, like this one.

He and his wife, Mary, now live high up on a ridge in remote Alaska, with moose and bear in the yard and eagles hunting from the roof.

Dedication

To Colleen Riley of Kachemak Research Development for helping me reach the pinnacle of becoming the Dummy of my dreams.

Author's Acknowledgments

Susan Pink is as much an author of this book as I am. She has the ability of taking a clunky, obscure, and badly worded thought and turning it into something that is easy to read. Whenever you come across something in this book that is clear and understandable, you can bet she had a hand in making it that way.

I would like to thank Melody Lane at Wiley Publishing for having faith that I could write this book. I would also like to thank Margot Maley Hutchinson at Waterside for helping convince Melody Lane that I could write the book.

Allen Wyatt contributed to the book by checking it for accuracy. I owe him a great thanks, but it's hard to be nice to some who uncovered so many of your mistakes.

Throughout the entire process, Jill Reitma at SPSS was very efficient and went to a great deal of trouble to make sure I had all the information and software I needed. Several people at SPSS made themselves available to me and answered even my silliest questions. The SPSS support group never left me in need of information.

I received valuable moral support from Jeanette Shafer, Brandon Wilson, and Garth Bradshaw, my co-workers at KRD, throughout the writing process. It was moral support — or they were simply laughing at me.

Publisher's Acknowledgments

We're proud of this book; please send us your comments through our online registration form located at www.dummies.com/register/.

Some of the people who helped bring this book to market include the following:

Acquisitions, Editorial, and Media Development

Project Editor: Susan Pink

Acquisitions Editor: Melody Layne

Technical Editor: Allen Wyatt, Discovery Computing, Inc.

Editorial Manager: Jodi Jensen

Media Development Specialists: Angela Denny, Kate Jenkins, Steven Kudirka, Kit Malone

Media Development Coordinator: Laura Atkinson

Media Project Supervisor: Laura Moss

Media Development Manager: Laura VanWinkle

Editorial Assistant: Amanda Foxworth

Sr. Editorial Assistant: Cherie Case

Cartoons: Rich Tennant (www.the5thwave.com)

Composition Services

Project Coordinator: Heather Kolter

Layout and Graphics: Carl Byers, Barbara Moore, Laura Pence, Ronald Terry

Proofreaders: Laura Albert, Aptara

Indexer: Aptara

Anniversary Logo Design: Richard Pacifico

Special Help: Laura Bowman

Publishing and Editorial for Technology Dummies

Richard Swadley, Vice President and Executive Group Publisher

Andy Cummings, Vice President and Publisher

Mary Bednarek, Executive Acquisitions Director

Mary C. Corder, Editorial Director

Publishing for Consumer Dummies

Diane Graves Steele, Vice President and Publisher

Joyce Pepple, Acquisitions Director

Composition Services

Gerry Fahey, Vice President of Production Services

Debbie Stailey, Director of Composition Services

Contents at a Glance

Table of Contents

· ·

Introduction

● ●

Good news! You don't have to know diddly-squat about statistics to be able to come up with well-calculated conclusions and display them in fancy graphs. All you need is the SPSS software and a bunch of numbers. This book shows you how to type the numbers and produce brilliant statistics. It really is as simple as that.

About This Book

This is fundamentally a reference book. Parts of the book are written as stand-alone tutorials to make it easy for you to get into whatever you're after. After you're up and running with SPSS, you can skip around and read just the sections you need. You really don't want to read straight through the entire book. That way leads to boredom. I know — I went straight through everything to write the book, and believe me, you don't want to do that.

The book was designed to be used as follows:

1. Read the opening chapter so you'll understand what SPSS is. I tried to leave out the boring parts.

2. If SPSS is not already installed, you may need to read about installing it.

3. Read the stuff in Chapter 4 about defining variables and entering data. It all makes sense after you get the hang of it, but the process seems to be kind of screwy until you see how it works.

4. Skip around to find the things you want to do.

I would mention that you could skip the introduction, but it's too late for that. Besides, you'll find some information here that could be useful.

This book is not about statistics. You will not find one explanation of statistical theory or how calculations are performed. This book is about what you can do to command SPSS to calculate statistics for you. The inside truth is that you can be as dumb as a post about statistical calculation techniques and still use SPSS to produce some nifty stats. You have my permission to stop thinking right now.

However, if you decide to study the techniques of statistical calculation, you'll be able to understand what SPSS does to produce numbers. Your main advantage in understanding the process to that degree of detail is that you'll be able to choose a calculation method that more closely models the reality you are trying to analyze — if you are interested in reality.

About the Data

Throughout the book you will find examples that use data stored in files. These files are freely available to you. The files are either installed with SPSS in the SPSS installation directory at `\Program Files\SPSS` (unless you chose another location during installation) or designed for this book and available on the following Web site:

`www.dummies.com/go/spss`

You can also find the files at my Web site:

`www.belugalake.com/spss`

Different kinds of files are available. Most are data files used to demonstrate statistical calculations, and some are programming source code files. (Would you believe you have a choice of three programming languages inside SPSS?)

Who This Book Is For

In general terms, this book is for anyone new to SPSS. No prior knowledge of statistics or mathematics is needed or even expected. In specific terms, this book was written with two groups in mind: students who are not majoring in mathematics but are instructed to use SPSS and office workers who are instructed to use SPSS.

For most people generating statistics, the complexity of using the software becomes an obstacle. My purpose in writing this book is to show you how to move that obstacle out of the way with minimum effort.

How This Book Is Organized

This book was written so you could read the first part, to get yourself started with SPSS, and then jump around to the other parts as needed. SPSS is a huge piece of software and you certainly don't want to use everything.

The book is filled with step-by-step procedures that you can follow to see how SPSS operates. After you use the provided sample data and step through an example, you will understand how to apply the example to your data.

The parts of the book divide the information about SPSS into its major categories. The chapters in each part further divide the information into smaller categories.

Part 1: The Fundamental Mechanics of SPSS

The first part is the only one intended to be read straight through. You can gloss over the installation, if you already have SPSS installed, but you will need to familiarize yourself with the configuration options. You will come across these configurations later and will need to know what can be changed. This is the only place in the book where you will find a complete example — starting with the entering of gathered data and ending with the generation of rudimentary analyses.

Part II: Getting Data into and out of SPSS

Input can be tricky. Variables are defined by type and size and a few other things. Part II shows you how to enter data through the main SPSS window or load it from a file. In fact, you can read data from several kinds of files. You can also write data to several kinds of files.

Part III: Graphing Data

In Part III you see how to produce graphs. A large part of the job performed by SPSS is displaying data in graphic formats. SPSS can produce lots of different kinds of graphs and maps. Fortunately, it's an easy thing to do — you simply select variable names and how you want them displayed.

Part IV: Analysis

Hidden down inside SPSS are lots of statistical methods. This thing manufactures numbers like McDonald's manufactures hamburgers. Part IV explains how to manufacture the numbers you want.

Part V: Programming SPSS with Command Syntax

Part V shows you how to use the SPSS internal command language. You can record procedures in Command Syntax and execute them at will. You can do anything with a Command Syntax program that you can do with the mouse and keyboard.

Part VI: Programming SPSS with Python and Scripts

Part VI is BASIC talk about programming and scripting SPSS. Anything you can do with Command Syntax or with the mouse and keyboard, you can also do in the Python programming language. The scripting language of SPSS is Sax BASIC.

Part VII: The Part of Tens

Part VII is all about the add-ons for SPSS and the locations on the Internet where you can find useful stuff.

Icons Used in This Book

You should remember this information. It is important to what you are doing.

Skip these unless the text makes you curious. This icon highlights unnecessary information, but I had to include it to complete the thought.

A tip highlights a point that can save you time and effort.

A warning is information about something that can sneak up and bite you.

Where to Go from Here

Read the first chapter. Then, if necessary, install SPSS, referring to Chapter 2. Work through the example in Chapter 3.

Now you're up and running. Figure out what you want to do and refer to the sections of the book necessary to do that. For some tasks (such as programming Python), you need to read an entire chapter. For other jobs, you need to read only a single section.

Part I
The Fundamental Mechanics of SPSS

The 5th Wave   By Rich Tennant

"The top line represents our revenue, the middle line is our inventory, and the bottom line shows the rate of my hair loss over the same period."

In this part . . .

This is a look at SPSS from 10,000 feet. Even if you know nothing whatsoever about SPSS, after you read this part you will have a good idea of how it all works. You won't know about all the details, but you will have a clear understanding of the general operation of SPSS. Everything else you find out about SPSS will fit in the structure you build for yourself by reading Part I.

This is the only part of the book intended to be read straight through. The only optional subject in Part 1 is the description of the installation if you've already installed SPSS.

Chapter 1

Introducing SPSS

• •

"There are three kinds of lies: lies, damn lies, and statistics." That statement is often attributed to Mark Twain, but that's not quite right. Mark Twain did say it, but he attributed it to someone else. He indirectly attributed it to Disraeli, but his attribution was vague, and the original statement, if it exists, can't be located. Speaking statistically, the odds are in favor of us never knowing who said it first.

Garbage In, Garbage Out

Statistical analysis is like a sewer. What you get out of it largely depends on what you put into it.

Over 82 percent of all statistics are made up on the spot to try to prove a point.

You can conclude just about anything if you're not careful with your data and with your calculations. SPSS watches the performance of the calculations for you, but the raw data, and which calculations should be performed, is up to you.

Let me show you a simple example of using raw data to produce an obviously wrong conclusion. Suppose you want to demonstrate, by sampling, that every odd number is prime. (A prime number can be evenly divided only by 1 and itself.) The first thing to do is gather a collection of data points, as shown in Table 1-1.

Table 1-1	Odd Numbers and Whether They Are Prime	
Number	*Prime?*	*Comment*
1	Yes	It fits the definition exactly
3	Yes	It is certainly both odd and prime
5	Yes	It fits the pattern of primes

(continued)

Table 1-1 *(continued)*		
Number	*Prime?*	*Comment*
7	Yes	So far, so good
9	No	Must be a bad data point, so throw it out
11	Yes	Now we're back on track
13	Yes	Looking good

Lots of things are already wrong with the data in Table 1-1. For one, the sample is too small. For another, the sampling cannot be considered random. All too often it happens that data points don't fit a preconceived conclusion, so they are omitted. The result of the data in this table can be used as proof of a fact that is dead wrong.

This book is not about the accuracy, correctness, or completeness of the input data. Your data is up to you. This book shows you how to take the numbers you already have, put them into SPSS, crunch them, and display the results so it all makes sense. Gathering valid data and figuring out which crunch to use is up to you.

From Whence SPSS?

SPSS is probably older than you are. In 2007 it becomes 38 years old, and the average age of an American is 35.3.

At Stanford University in the late 1960s, Norman H. Nie, C. Hadlai (Tex) Hull, and Dale H. Bent developed the original software system named Statistical Package for the Social Sciences (SPSS). They needed to analyze a large volume of social science data, so they wrote software to do it. The software package caught on with other folks at universities and, with the open source tradition of the day, the software spread through universities around the country.

The three men produced a manual in the 1970s and the software's popularity took off. A version of it existed for each of the different kinds of mainframe computers of the time. Its popularity spread from universities into other areas of government, and it began to leak out into private enterprise.

In the 1980s, a version of the software was moved to the personal computer, and here we are today.

Maybe it has been continuously successful because the software does such a good job of making predictions, and the SPSS people could always figure out what they should do next.

The Four Ways to Talk to SPSS

More than one way exists for you to command SPSS to do your bidding. And you don't have to choose one and stick with it — you can perform tasks using whichever of the four interfaces you prefer. You can use any of the four approaches to perform any of the SPSS functions, but which one is best for you depends, to an extent, on the task to be performed and which interface you prefer:

- ✔ **GUI (graphic user interface):** SPSS has a windowing interface and commands can be issued by the mouse through menu selections that cause dialog boxes to appear. This is a fill-in-the-blanks approach to statistical analysis that guides you through the process of making choices and selecting values. The advantage of the GUI approach is that, at each step, SPSS will make sure that you enter everything necessary before proceeding to the next step. This is the preferred interface for those just starting out — and if you don't do much with SPSS, this may be the only interface you ever use.

- ✔ **Syntax:** This is the internal language used to command actions from SPSS. It was known as the command syntax of SPSS, hence its name. It is often referred to as the command language. You can write Syntax commands to directly command SPSS to do anything it is capable of doing. In fact, when you use menu and dialog box selections to command SPSS, you are actually generating Syntax commands internally that do your bidding. That is, the GUI is nothing more than the front end of a Syntax command writing utility. Writing (and saving) command language programs is a good way to store processes that you expect to repeat. You can even grab a copy of the Syntax commands generated by the GUI and save them to be repeated later.

- ✔ **Python:** This is a general-purpose language that has a collection of SPSS modules written for it, making it possible to write programs that work inside SPSS. It can be run with the Syntax language to command SPSS to perform statistical functions. One advantage of using Python is the fact that it is a modern language and gives you the power and convenience that come with languages today, including the ability to construct a more readable program. In addition, because it's a general-purpose language, you can read and write data from other applications and from other files.

- ✔ **Scripts:** The items that SPSS calls scripts are actually programs written in BASIC. This language is simple and many people are familiar with it. Also, a BASIC program can be written as an *autoscript* — a script that executes automatically when SPSS produces certain output.

The Things You Can and Cannot Do with SPSS

The full-blown SPSS package comes in many parts. The Base system is the center around which the rest of SPSS revolves. You have a Base system. You may also have one or more add-ons. With only one exception, everything described in this book is included in the Base system, so you will be able to do anything you read about. The one exception is the Python programming language, which requires some additional software. But the software is a free download and also comes on the SPSS distribution CD. Chapter 20 describes other modules you can add to your Base system.

SPSS works with numbers. Only. If you cannot express your information as a number, you can't run it through SPSS. You will see names and descriptions seemingly being processed by SPSS, but that's because each name has been assigned a number. That's why survey questions are written like, "How much do you enjoy eating rhubarb? Select your answer: Very much, sort of, don't care, not really, I hate the stuff." A number is assigned to each of the possible answers, and these numbers are fed through the statistical process. SPSS uses the numbers, not the words, so be careful about keeping all your words and numbers straight.

You must keep accurate records describing your data, how you got the data, and what it means. SPSS can do all the calculations for you, but only you can decipher what it means. In *Hitchhiker's Guide to the Galaxy,* a computer the size of a planet crunched on a problem for generations and finally came out with the answer, 42. But the people tending the machine had no idea what the answer meant because they didn't remember the question. They hadn't kept track of their input. You must keep careful track of your data or you may later discover, for example, that what you have interpreted to be a simple increase is actually an increase in your rate of decrease. Oops.

SPSS lets you enter the data and tag it to help keep it organized, but you already have the data written down someplace and fully annotated. Don't you?

How SPSS Works

The developers of SPSS have made every effort to make the software easy to use. This prevents you from making mistakes or even forgetting something. That's not to say it's not possible to do something wrong, but the SPSS software works hard to keep you from running into the ditch. To foul things up, you almost have to work at figuring out a way of doing something wrong.

You always begin by defining a set of *variables,* then you enter data for the variables to create a number of *cases.* For example, if you are doing an analysis of automobiles, each car in your study would be a case. The variables that define the cases could be things such as the year of manufacture, horsepower, and cubic inches of displacement. Each car in the study is defined as a single case, and each case is defined as a set of values assigned to the collection of variables. Every case has a value for each variable. (Well, you can have a missing value, but that's a special situation described later.)

Variables have types. That is, each variable is defined as containing a specific kind of number. For example, a *scale* variable is a numeric measurement, such as weight or miles per gallon. A *categorical* variable contains values that define a category; for example, a variable named gender could be a categorical variable defined to contain only values 1 for female and 2 for male. Things that make sense for one type of variable don't necessarily make sense for another. For example, it makes sense to calculate the average miles per gallon, but not the average gender.

After your data is entered into SPSS — your cases are all defined by values stored in the variables — you can run an analysis. You have already finished the hard part. Running an analysis on the data is much easier than entering the data. To run an analysis, you select the one you want to run from the menu, select appropriate variables, and click the OK button. SPSS reads through all your cases, performs the analysis, and presents you with the output.

You can instruct SPSS to draw graphs and charts the same way you instruct it to do an analysis. You select the desired graph from the menu, assign variables to it, and click OK.

When preparing SPSS to run an analysis or draw a graph, the OK button is unavailable until you have made all the choices necessary to produce output. Not only does SPSS require that you select a sufficient number of variables to produce output, it also requires that you choose the right kinds of variables. If a categorical variable is required for a certain slot, SPSS will not allow you to choose any other kind. Whether the output makes sense is up to you and your data, but SPSS makes certain that the choices you make can be used to produce some kind of result.

All output from SPSS goes to the same place — a dialog box named SPSS Viewer. It opens to display the results of whatever you've done. After you have output, if you perform some action that produces more output, the new output is displayed in the same dialog box. And almost anything you do produces output.

All the Strange Words

Statistics seems to have been born in the land of strange words. Lots of them. If you come across a term that you don't understand, such as *dichotomy, variable,* or *kurtosis,* you can look it up in the glossary at the back of the book.

It's not only new words that can trip you up. You will find common words used in a special way. For example, a *break variable* has a special purpose when organizing tabular data.

The glossary is always there, ready to explain the meaning of those strange terms.

All Those Files

Input data and statistics are stored in files. Different kinds of files. Some files contain numbers and definitions of numbers. Some files contain graphics. Some files contain both.

The examples in this book require the use of files that contain data configured to demonstrate capabilities of SPSS. These files are all in one of two places. Most are in the same directory you use to install SPSS. That is, the action of installing SPSS also installs a number of data files ready to be loaded into SPSS and used for analysis. A few of the files used in the examples can be found in the compressed file `spss.zip` found at this Web site:

```
www.dummies.com/go/spss
```

You can also get the files from the author's Web site:

```
www.belugalake.com/spss
```

After you have downloaded the zip file to your system, you need to decompress (unzip) it into separate files and directories. If you don't have an unzipper, and would like to get one, enter the search word *unzip* into Google. There are free ones and commercial ones. A popular commercial product that runs as a windowing program and uses mouse controls can be found at the following Web site:

```
http://www.winzip.com
```

If you don't want to buy WinZip, you can download a free trial version that will work just fine for this job.

If you want, the Web site is configured so that you can download the files in the form ready to be used. Doing it this way is a bit easier, but only if you don't need all the files. Each file will have to be downloaded individually.

Where to Get Help When You Need It

You're not alone. Some immediate help comes directly from the SPSS software package, and other help can be found on the Internet. If you find yourself stumped on some point, you can look in several places:

- ✔ **Topics:** Choosing Help➪Topics from the main window of the SPSS application is your gateway to immediate help. The help is somewhat terse, but it will often be exactly what you need. You will find all the information in one large help document, presented to you one page at a time. Choose Contents to select a heading from an extensive table of contents, choose Index to search for a heading by entering its name, or choose Search to enter a string search inside the body of the help text.

  In the help directory, the titles in all uppercase are descriptions of Syntax language commands.

- ✔ **Tutorial:** Choose Help➪Tutorial to open a dialog box with the outline of a tutorial that guides you through many parts of SPSS. You can start at the beginning and view each lesson in turn, or you can select your subject and view just that.

- ✔ **Case Studies:** Choose Help➪Case Studies to open a dialog box containing examples in a format similar to that of the Tutorial selection. You can select titles from its outline and view descriptions and examples of specific instances of using SPSS. You will also find descriptions of the different types of calculations. This is a good place to look if some particular analysis type is eluding your comprehension.

- ✔ **Statistics Coach:** Choose Help➪Statistics Coach if you have a good idea of what you want to do but need some specific information on how to go about doing it.

- ✔ **Command Syntax Reference:** Choose Help➪Command Syntax Reference to display more than 2000 pages of references to the Syntax language in your PDF viewer. The regular help topics, mentioned previously, provide a brief overview of each topic, but this document is much more detailed.

- ✔ **Python:** Choose Help➪Programmability to display a 100-page PDF document on programming SPSS using Python.

Your Most Valuable Possession

The most valuable possession you have in dealing with statistics is not your computer. It's not your SPSS software. It's not this book, or any other book you may be using to learn statistics. You can lose any one of those, but any one of them can be replaced.

Your most valuable possession is your data. Sure, you can always go and get more data, but you can't go and get the same data. The world doesn't hold still long enough. Make sure you make backup copies of your data.

Back up your data to memory that does not live in the same building with the computer you are using. You can swap backups with a friend, or if you have access to a remote Web site, you can stuff files in a blind directory.

This message about backing up your data comes to you from someone who has been stung. Twice. And I don't want to talk about it again. Ever.

You Can Dive As Deep As You Want to Go

SPSS makes no effort to keep anything a secret. It is designed to be as easy to use as possible, so you really don't have to know that much to make it work. However, if you want to understand how things are working internally, you can find out if you dig. And you don't have to dig very far. Choosing Help is the first step to finding out anything you want to know about what's going on inside.

Let's say you are working on your numbers and want to use some specific algorithm to do your calculations. SPSS has been at this longer than you have, so the algorithm you want to use is almost certainly built in. If you are not sure exactly what SPSS is doing to calculate some of the numbers, you can go to the Help menu and the PDF documents on the documentation CD to find out how the calculations are being performed. But, before you start looking, make sure you really want to know because the equations and how they are applied are explained in excruciating detail.

The purpose of this book is to give the shallow divers enough information to be able to swim, and show the deeper divers how to begin. I don't explain all the details because there are too many. There's simply not enough room in a book this size to explain SPSS in depth.

Chapter 2

Installing and Running the Software

*T*his chapter is all about installing your software and setting the options that determine how it works. If the software you'll be using is already installed, you can skip the first part of this chapter. That is, don't install the software if it's already installed. I mention that only because this is a *For Dummies* book and I was told not to leave anything out.

The installation process guides you, step by step, and then does most of the work itself. The configuration settings all default to something reasonable, so the only ones you might want to change are the ones you have some gripe with. I suggest leaving them alone for now.

Getting SPSS into Your Computer

Soap powder comes in boxes, paint comes in cans, corn dogs come on sticks, and SPSS comes on CDs. Two CDs. Open the package in which it came and read the labels on the CDs, and you'll see that one contains the SPSS software and the other contains documentation. The CD with the software also contains some other stuff, which is listed on the CD label, but you can ignore that for right now.

Find a place to put the package and all its contents. Don't throw out anything. That includes the plastic box in which you found the CD and the cardboard stiffener that came inside the mailing package. Trust me, you'll need them later. Use a folder, designate a drawer, or clear a spot on a shelf — keep everything in one place so you can find it later.

The Mac version of the software is similar, but details of the installation procedure described here are specific to Windows.

The things you need

You won't have to worry about the minimum requirements for the computer — unless yours is an antique. I mean, who doesn't have at least 256MB of RAM and 300MB of free disk space?

SPSS comes in a variety of flavors. They're fundamentally alike, but some versions have more parts than others. You may have all, some, or none of the add-ons described in Chapter 20. In any case, you need an authorization code to enable whatever you do have. You may have more than one authorization code — it depends on how your SPSS system is configured, which is determined by what parts are included with it.

Remember those bits of paper I warned you to keep track of? You will find your authorization code, or codes, somewhere there. Go ahead and find them now.

For the installation procedure to work, you must be logged in to your Windows system with administrator privileges. You don't have to be logged in as an administrator, but whatever login you are using must have the privileges that the administrator has.

You should also be connected to the Internet. You can install SPSS without being connected, but it's a pain to do it that way. Make it easy on yourself and connect your computer to the Internet before you start. And keep it connected at least until you get SPSS installed.

In summary, before you begin the installation:

- ✔ You must have access to your authorization code or codes.
- ✔ You must have access to the serial number of your copy of SPSS.
- ✔ You may also need to have access to your customer number.
- ✔ You must be logged into your computer with administrator privileges.
- ✔ For convenience, you probably want to be connected to the Internet.

Cranking up the installer

The installation procedure is dead simple. You simply start the installation program and answer the questions. And the questions are easy.

You can start the installer in two ways. The first method is automatic: You insert the SPSS CD into the drive and wait a bit. Most Windows computers will recognize what's on the CD and start the installer. If the installer doesn't start automatically (or if you fool around and close the window after it started), choose Start⇨Run and execute the program on the CD named `setup.exe`. Either way, you get the window shown in Figure 2-1.

As you can see in Figure 2-1, it's possible to install several items, some of which you may have never even heard of. Stick with me and you'll hear about all of them eventually. For starters, however, we'll look at the simple case of a single-user installation of SPSS.

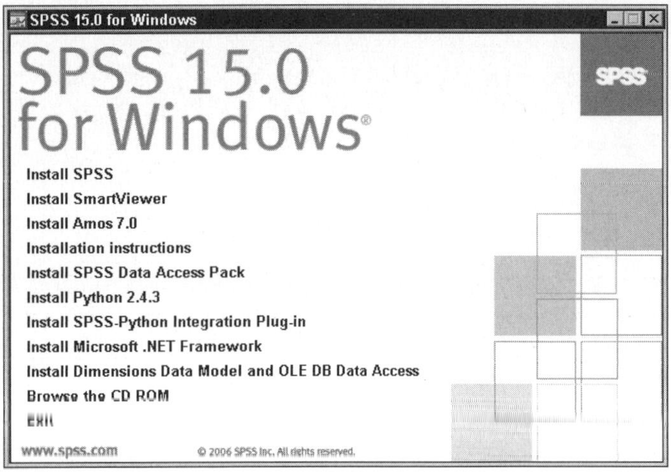

Figure 2-1:
The first window is a list of installation choices.

The SPSS installation sequence

With the window shown in Figure 2-1 on your screen, click the words Install SPSS, at the very top of the list. The computer makes a sort of boink sound (if speakers are attached) and you are informed that something called InstallShield is preparing itself. While InstallShield is getting ready, it displays some animation on the screen to indicate that progress is being made. The software that will install SPSS is getting *itself* installed. When it's finally satisfied that everything is okay for the installation to proceed, it pops up the window shown in Figure 2-2. We'll be installing the software for a single user, so choose the Single User License option and then click Next. If you need to make an installation for multiple users, arrangements would have been made at the time of the purchase of SPSS and your administrator would have supplied you with the necessary information. If you *are* the administrator and you don't know what you should do, you need to get in touch with SPSS. The instructions in this book are for a simple installation.

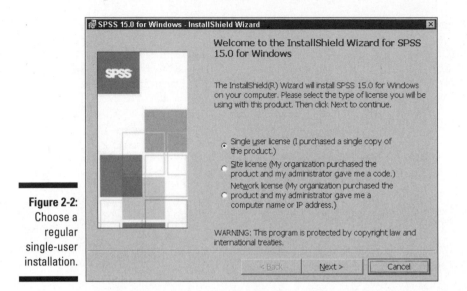

Figure 2-2:
Choose a
regular
single-user
installation.

After you make your selection, you're greeted by the license, as shown in Figure 2-3. Simply do what it says: Read the license and if you accept the terms, select the I Accept the Terms in the License Agreement option and then click the Next button.

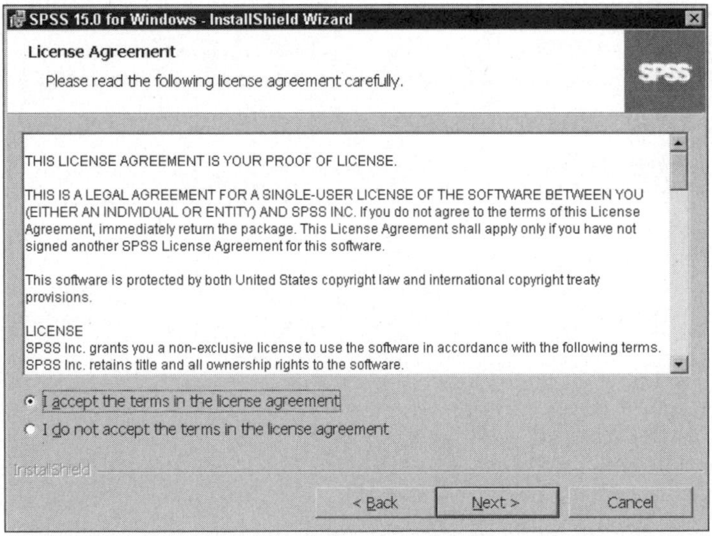

Figure 2-3:
The license
agreement
you have
with SPSS.

Now the installation gets talky, as shown in Figure 2-4. If you thought the license was something, wait until you read this stuff. Not all of it will apply to you, but you should read it anyway because you might find something you need, such as the Klingon numbering system. After you've gained all the pleasure you can stand from the ReadMe file, click Next to move on.

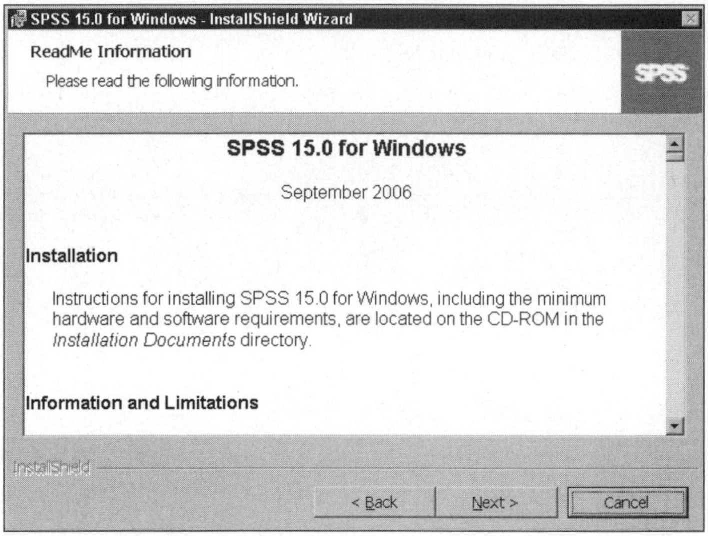

Figure 2-4:
The
ReadMe
notes
contain the
last-minute,
"Oh, by
the way"
thoughts.

The next screen, shown in Figure 2-5, asks for your name and organization. I always take it as a compliment that the software thinks of me as being organized, but I can never figure out what to put in the blank. You can put anything you like in there, but keep it clean because it could pop up on the screen one afternoon while your mom is watching. The third piece of information is a little more important. It wants you to enter the serial number of your copy of the software. This is *not* the authorization code — that comes later. You can find the serial number in two places: on a tag inside the plastic box in which you found the CD and on the cardboard stiffener that came inside the mailing package. See, I told you not to throw anything away.

When you click the Next button, you get the window shown in Figure 2-6, which asks for the directory into which SPSS will be installed. The directory it chooses is fine, and you should change it only if you have a really good reason. If you can't think of a reason, accept what's there and move on by clicking the Next button.

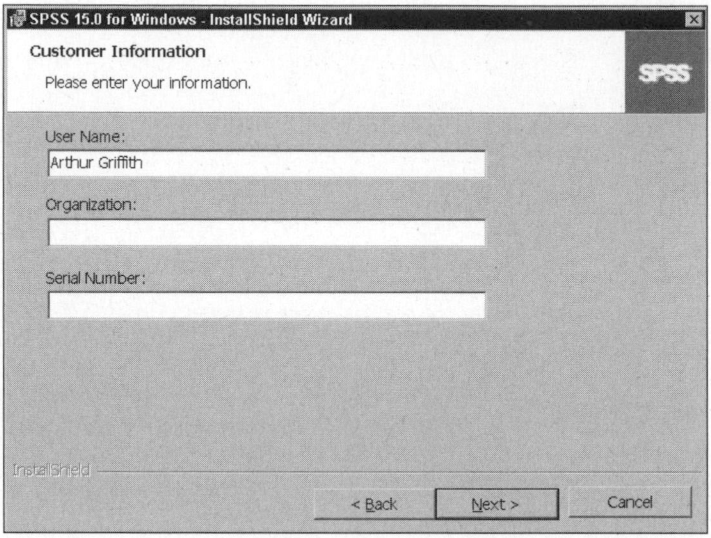

Figure 2-5:
Name, organization, and the serial numbers.

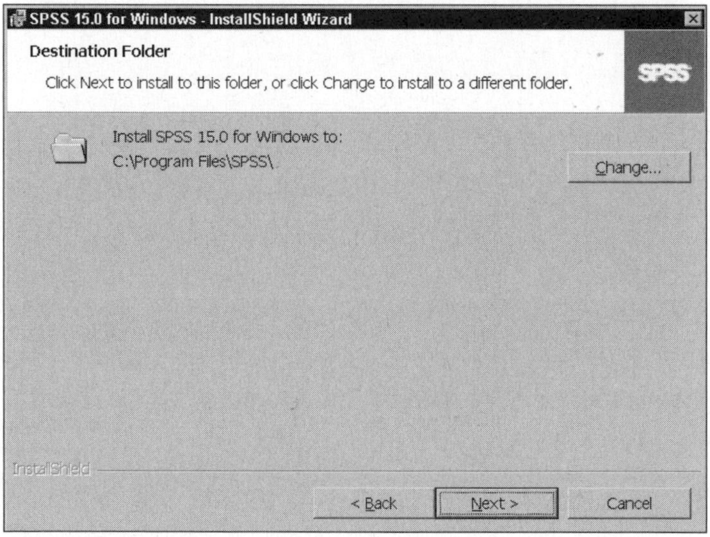

Figure 2-6:
The directory into which SPSS will be installed.

This displays a window that asks you whether you want to install SPSS. All you've done so far is answer some questions; nothing has been installed. This window has a Back button you can use to go back and change your answers. The Next button unleashes the installation software onto your computer. The screen also has a Cancel button if you chicken out or if you enjoyed the process so much that you want to drop everything and do the entire thing over again. You want SPSS on your computer, so click Next.

The next window, shown in Figure 2-7, lists every file being installed, while a progress indicator moves across the screen. The file names flicker by pretty fast; only Superman or Data from Star Trek can read them. Normal mortals see mostly a line of constantly flickering letters.

The progress indicator marches across the screen until it reaches the far right. At that point, the flickering of file names will stop. For a time, nothing moves. Be patient. Just about the time you start to wonder whether something has gone wrong, the display presents the window shown in Figure 2-8.

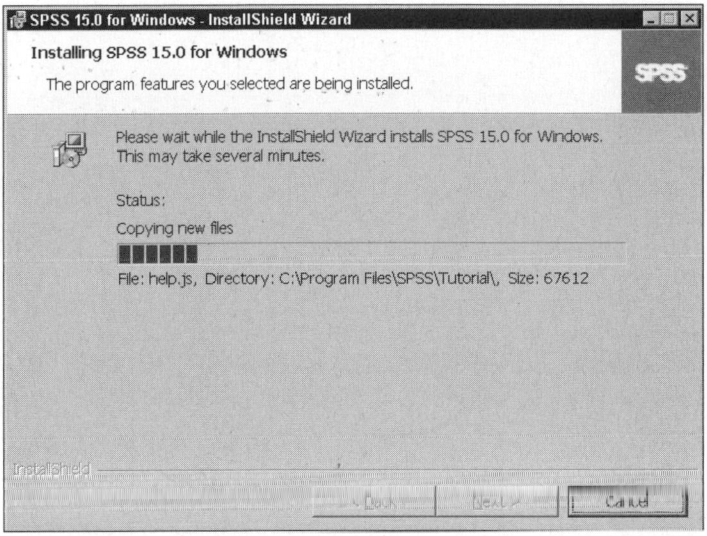

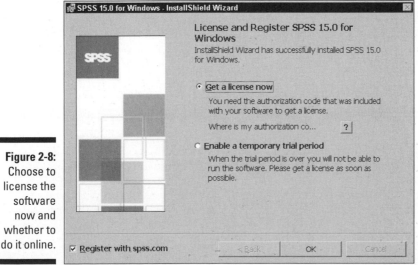

Figure 2-7:
An animated window reassures you that something is happening.

Figure 2-8:
Choose to license the software now and whether to do it online.

This is where it is convenient to be connected to the Internet. You need to select both options shown in Figure 2-8 if you want to enable your copy of SPSS. If you click the question mark button, you are told that you have the authorization code on one of your pieces of paper or maybe someone else has it. If you don't have it, you might as well stop until you get it because the following steps require it. If you already have it, click OK and you will get the startup window of the License Authorization Wizard, shown in Figure 2-9.

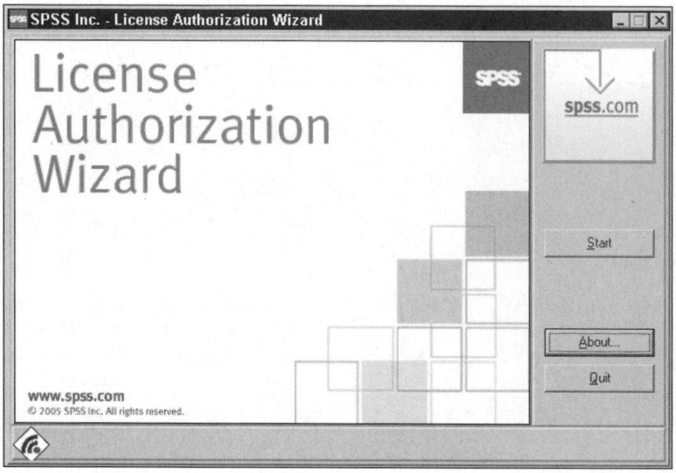

Figure 2-9:
The License
Authorization
Wizard.

If you want, you can read about the License Authorization Wizard by clicking the About button in Figure 2-9, but that won't put you any closer to getting the software ready to run. To move forward with the installation, click Start.

Your software is installed, but you can't use it because you don't have a license. You need to get your license from the SPSS company, and to do that you need the authorization code that came in your software package. As you can see by the window in Figure 2-10, you have four ways of getting a license. You want the first choice — via the Internet.

Click Next, and the window shown in Figure 2-11 appears, asking for your authorization code. This is your big moment. Enter your authorization code exactly as it is printed on your piece of paper, and then click Next. If you enter the right code and are doing your licensing through the Internet, you are licensed in just a few seconds. The window changes to tell you that you're doing it correctly, and a Finish button appears at the bottom. Click Finish and you're out of there. You now have SPSS installed and ready.

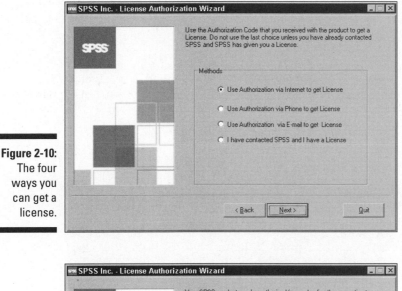

Figure 2-10:
The four
ways you
can get a
license.

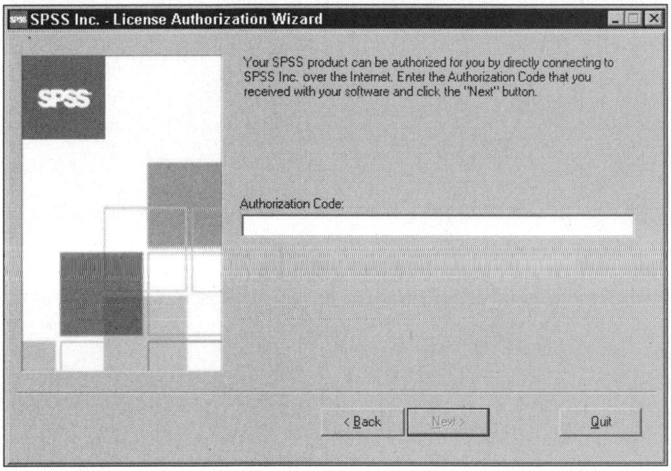

Figure 2-11:
The prompt
for your
authorization
code.

Registration

Now that SPSS is on your system and ready to run, that's it, right? Not so fast. You need to register. When you finished your authorization, a Web site popped up displaying your contact information. This Web site enables you to correct and complete your information, so read it over carefully. It's to your advantage to get the information right. For one, you can use the Web site to add your address to some useful mailing lists.

You may need one more number. It could be that your customer number is not recorded on the Web site, and if that's the case you can enter it now. You will find it on one of those pieces of paper I told you to save — probably the same one that has your authorization code.

The Internet being the Internet, your connection might get dropped right in the middle of getting registered. If that happens, you can get back in by going to the following Web site: http://www.spss.com/registration.

You have to use your e-mail address and a password to get back in. If you don't know your password, you can have it e-mailed to you. People say you should memorize your password instead of writing it down, but trying to do that always gets me in trouble. I just write it on one of the pieces of paper I stash with the other SPSS stuff. It's okay, though, because nobody in the world can read my handwriting.

Starting SPSS

You now have SPSS installed on your computer. You'll find a listing for it with the other programs on your Start menu. Choose Start➪Programs➪SPSS for Windows. You then have three choices:

- ✔ SPSS for Windows
- ✔ License Authorization Wizard
- ✔ Production Mode Facility

The first choice is the main program itself and will be the number-one selection on your hit parade in days to come. The second choice is the authorization stuff you went through earlier. The third choice allows you to preset instructions so you can run programs while you are off doing something else. For example, if you have an analysis you run every day to produce a report, you can set the analysis to run automatically. Later chapters go into all this. For now, let's stick to the main activity, choice number one.

When you first start SPSS, you get a window like the one in Figure 2-12. This window makes it possible for you to go directly to the window you want to work with. The problem is that it assumes you already know what you want to do, but so far you have no idea what you want to do, so just click the Cancel button to close the window.

You will see the regular Data Editor window, shown in Figure 2-13. If you've ever worked with a spreadsheet, this screen should look familiar. And it works much the same way. This window is the one you use to enter data. I generally like to expand the window to fill the entire screen because more spaces are displayed at one time. Besides, I don't need to see any other windows because I almost never do two things at once.

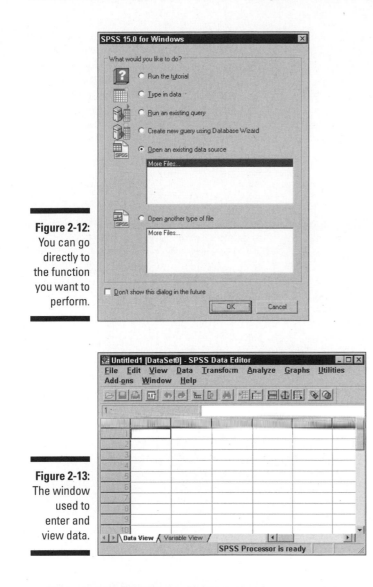

Figure 2-12:
You can go
directly to
the function
you want to
perform.

Figure 2-13:
The window
used to
enter and
view data.

The Default Settings and What They Can Become

Over time, you'll find that you want to configure your system to work in ways
you prefer. SPSS has lots of options that you can set to do just that. If you are
new to this and have just started looking at the software, you probably don't
want to change many options just yet, but you need to have some idea of
what they are and what you can do with them. Later, when you absolutely
have to make some sort of change, you will know where to go to do it.

With the Data Editor on the screen (refer to Figure 2-13), choose Edit⟿ Options to display the Options window. All possible options can be set in the Options window. At the top of this window are some tabs, and each tab selects a different collection of options. Sometimes a change in configuration doesn't have an immediate effect. For example, if you change the way values are labeled in a report that's already displayed, nothing happens because the report has already been constructed. You have to run the report-generating software again to have the changes take effect.

General options

The first tab in the Options window, the General tab, displays a dialog box with options that don't fit into any of the categories defined by the other tabs. This tab is shown in Figure 2-14.

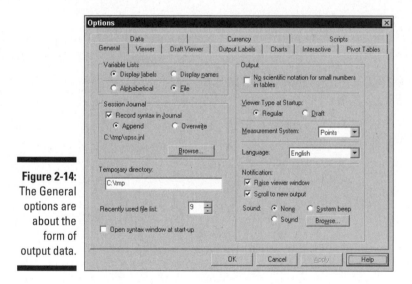

Figure 2-14:
The General options are about the form of output data.

The options displayed by the General tab follow:

> ✔ **Variable Lists:** Lists of variables in your output can be identified by either their labels or their names. You can think of these as short titles and long titles, and you can have your data, by default, tagged by one or the other as they appear in lists. Also, you can have your data appear in alphabetical order by the title you use for it or simply by the order in which the data appears in the file. File order usually makes more sense.

✔ **Session Journal:** Records a list of all commands executed on the data, in the order they are executed. You can select whether you want to erase the old journal file and start a new one each time and simply over-write the old one, or whether you just want to continuously append the new text onto the end of one long journal. You can also select the directory and file name to become the journal.

✔ **Temporary Directory:** A workspace for SPSS. If you decide to change it, choose a place containing only files you don't need.

✔ **Recently Used File List:** The files you've recently opened. The number specifies how many are listed on the menu.

✔ **Open the Syntax Window at Start-up:** Makes SPSS begin with the syntax window instead of the data editor. Choose this option if you use the scripting language more often than the windowing interface to enter data and run your predefined procedures.

✔ **Output:** Suppresses scientific notation for small numbers. For example, 12 appears as 12 instead of 1.2e1, which is a little harder to read. SPSS doesn't say exactly what it considers to be a small number.

✔ **Viewer Type at Startup:** The viewer to generate at startup. In general, Regular produces a better layout for interactive displays and Draft is more suitable for output that will be written to a file.

✔ **Measurement System:** Units used to specify the margins between table cells, the width of cells, and the spacing between printed characters. You can use inches, centimeters, or the default, points. (A point is $\frac{1}{72}$ of an inch.)

✔ **Language:** Set to any one of about a dozen choices. It makes life easier if you choose a language you actually know how to read.

✔ **Notification:** The method the software uses to notify you when the results of a calculation are available. With the Raise Viewer Window option, the display window opens automatically. With the Scroll to New Output option, the window scrolls and exposes the location of the new data. You also can have the system beep, tweet, or sing when an analysis is complete. It's considered impolite to have it make rude comments when an analysis finishes.

Viewer options

Output from SPSS is formatted for viewing with either the draft viewer or the regular viewer. SPSS thinks in terms of a printed page, but the same layouts are used for displaying data on the screen. The options you can set for the regular viewer can be accessed with the Viewer tab, shown in Figure 2-15.

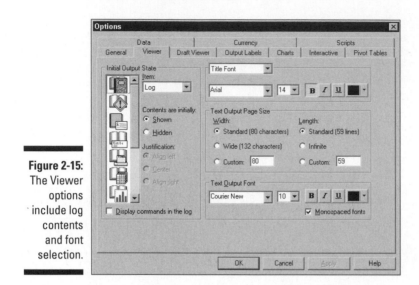

Figure 2-15:
The Viewer
options
include log
contents
and font
selection.

The options in the Viewer tab follow:

- **Initial Output State:** Determines which items are displayed each time you run a procedure. You choose an item by either selecting its name (Log, Warnings, Notes, Title, and so on) from the pull-down list or by selecting its icon. Then you can select whether you want it to appear or remain hidden, how you want its text justified (Align Left, Centered, or Align Right), and whether the information occurrence should be included as part of the log (Display Commands in the Log).

- **Title Font:** The font used for main output titles. It appears at the top of the first page of a report.

- **Page Title Font:** The font used for the title appearing at the top of subsequent pages of a report.

- **Text Output Page Size:** Determined by your printer. The settings are not obvious with most printers, and you may have to experiment with the Width and Length options to get page size just right. It's not your fault, it's just the way printers are.

- **Text Output Font:** Determines the font used for the text of your report and labeling on graphs and tables. The font size will also have some effect on the page width and length because the sizes are measured in a count of characters. Note: Some fonts have variable-width characters, which will cause your columns not to align correctly. If you want everything to always align in neat columns, use a monospaced font.

Draft viewer options

The draft viewer is a different and generally simpler format for the output of the analysis programs you run. The Draft Viewer options can be set using the window shown in Figure 2-16.

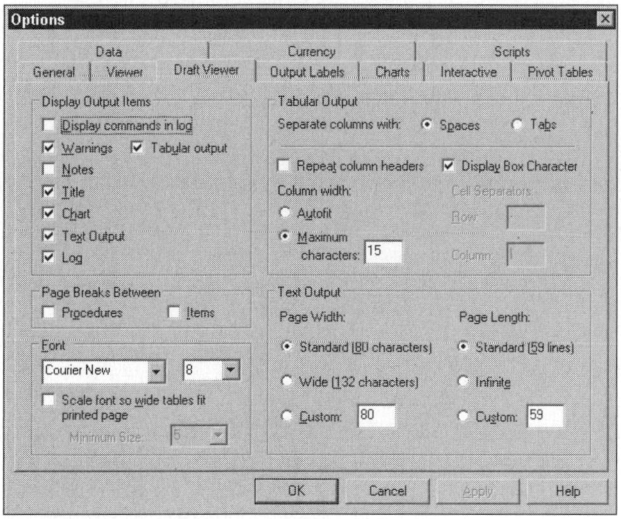

Figure 2-16: The window used to set the format of the output of the draft viewer.

Following are the options in the Draft Viewer tab:

- ✓ **Display Output Items:** Determines what is included in draft output. You can have all the commands written to the log, which is how you can have a series of syntax language commands saved so you can copy them later and include them in your own scripts. Sneaky, eh?

- ✓ **Page Breaks Between:** Inserts page breaks between procedures and between items.

- ✓ **Font:** Sets the size and type of the font. You can also force the font size to be changed automatically to make the output fit on a page, but when you do that, you also have to specify a maximum number of columns.

- ✓ **Tabular Output:** Uses tabs or spaces to separate columns in tables. Some printers work swell with one but not so good with the other. If tabs work, you may not need to use a monospaced font to line up the columns — but tab spacing can cause other things to get wacky.

  If you use spaces to position your columns, line wrapping doesn't happen and each column is set to the width of the longest string of characters that will fit in the column. But there is a maximum number of characters you can set for your columns. You'll need to experiment to see what happens. How's your paper budget?

Don't expect tabbed columns to align on the display. But laying out formats with tabs can be handy for copying the values for pasting into another application.

✔ **Repeat Column Headers:** Repeats column headers at the top of each page.

✔ **Display Box Character:** Inserts box characters — a graphic square that can't be mistaken for any number or letter — as separators between rows and columns. Without this option set, a vertical bar (|) separates columns and a minus sign (–) separates rows.

✔ **Column Width:** Sets the column width. You can do so only if you have selected Spaces to position the columns. If you specify Autofit, the column width is set to that of the longest label. If you specify a width, labels might wrap to another line.

✔ **Text Output:** Determines the size and whatnot of each page, but, here again, you have to experiment to see what happens.

Labeling output

Every variable can be identified in two ways: by a label and by a name. In your output, you can specify to have variables identified by one or the other or both. Output labeling is configured using the Output Labels tab, shown in Figure 2-17.

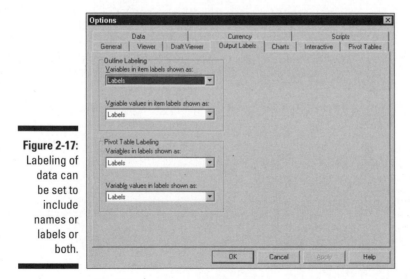

Figure 2-17: Labeling of data can be set to include names or labels or both.

You can display the variable names, the variable labels, or both, along with the values for the variables. Longer labels can be descriptive and make your data easier to determine, but they can also screw up some formats. Following are the options in the Output Labels tab:

✓ **Outline Labeling:** The text used to identify the parts of charts and graphs.

✓ **Pivot Table Labeling:** The text used to identify the rows and columns of tables.

Chart options

The default appearance of charts is determined by the settings in the Charts tab, shown in Figure 2-18.

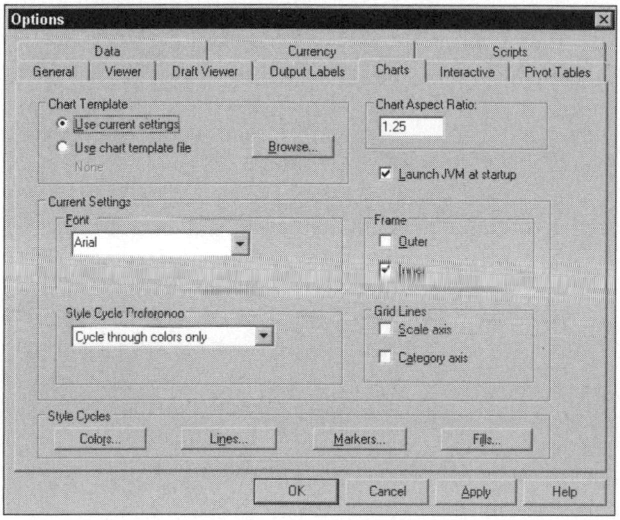

Figure 2-18:
Change the default appearance of a chart with these settings.

The options in the Chart tab follow:

✓ **Chart Template:** A file that contains a set of starter settings that you can use for designing a new chart. When you create a new chart, it can use the settings in this configuration window, or it can use this file. You can select any file to be your default starting template. It's easy to create a chart template: Simply create a chart that has all the configuration settings you like and save it so it can be used as the template file.

✓ **Font:** The default font for the text in any chart you design.

✔ **Style Cycle Preference:** How SPSS chooses the styles and colors when laying out data items in a chart. You can have SPSS cycle through colors only, which means that each item included in the graph is identified by its color. With a black and white printer or display, choose Cycle Through Patterns Only, in which each data item is identified by a graphic pattern of line styles and marker symbols.

✔ **Style Cycles:** Customizes the sequence of colors and patterns to be cycled through.

✔ **Chart Aspect Ratio:** The ratio of the width to the height of the produced charts, initially set to 1.25. It's a matter of opinion what ratio looks better. This is another place where you will have to experiment.

✔ **Launch JVM at Startup:** Starts the Java Virtual Machine when SPSS starts. SPSS starts more quickly if you turn off this option, but because some chart features use Java, the first chart you display using one of these features will take the delay hit.

✔ **Frame:** Determines whether charts display an inner frame, an outer frame, both, or neither.

✔ **Grid Lines:** Displays dividing lines on the scale axis, on the category axis, or on both.

Interactive chart options

An *interactive chart* can be embedded in another application as an ActiveX component. Some configurations apply only to interactive charts. These items can be set in the Interactive tab, shown in Figure 2-19.

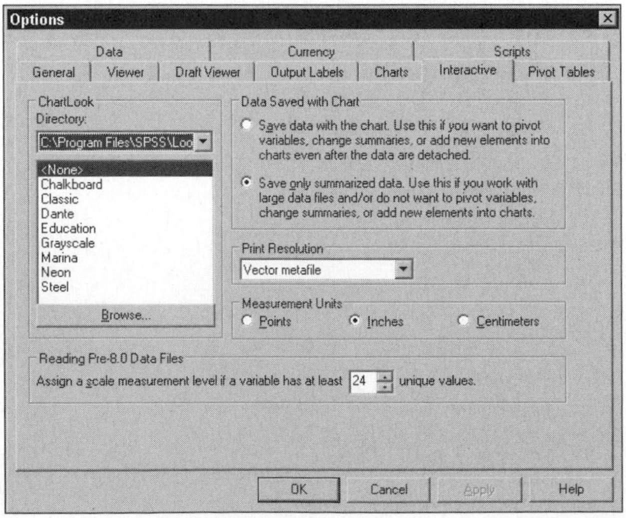

Figure 2-19: Settings for interactive charts.

The options in the Interactive tab follow:

- **ChartLook:** For the construction of interactive charts. It is similar to a chart template, which is for static charts. You select the file that you want to use as the basis for the appearance of your new charts. Several predefined files come with the system and are included in the list in this window. You also can create your own file by choosing ChartLook from the menu of the interactive graphics editor and saving the file under a name you load in later as the default.

- **Data Saved with Chart:** Specifies whether you want the data saved with the chart and how. Saving the data with the chart makes it possible for you to reformat the data even after you've separated it from the data files. With the chart separated from the data, you can add new data fields and such, but you can include only data present when you saved the chart.

- **Print Resolution:** Prepares an image for the printer. The High Resolution selection looks better but takes longer to print than Low Resolution. The Vector Metafile selection is fairly rapid and produces good results.

- **Measurement Units:** Used to lay out items on the printed chart.

- **Reading Pre-8.0 Data Files:** Necessary for reading data files produced by old versions of SPSS. You can have the read process assign a scale measurement level to any variable having at least the specified number of variables.

Pivot table options

The tabular output format of SPSS is the *pivot table*. An example is shown in Figure 2-20, which is the Pivot Tables tab used to set display options for the tables.

The options in the Pivot Tables tab follow:

- **TableLook:** A file that contains your standard pivot table and determines the initial appearance of any new tables you create. Several such files come with the system and are listed in the window. You can also create your own file by choosing TableLook from the menu in the pivot table editor window.

- **Set TableLook Directory:** Sets the currently displayed directory as the one in which your new table files are stored. You can choose any directory you like; clicking this button will cause your chosen directory to appear in this window by default.

✔ **Adjust Column Widths For:** Controls the way SPSS adjusts column widths in pivot tables. They can be adjusted according to the width of the labels or according to the width of the data or labels, whichever is wider.

✔ **Default Editing Mode:** Double-clicking a pivot table enables it for editing. This option determines whether that editing will be performed in place or in a separate window opened for the sole purpose of editing.

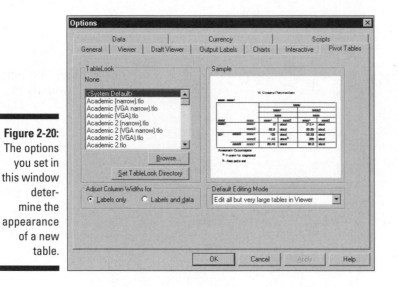

Figure 2-20:
The options you set in this window determine the appearance of a new table.

A few data-handling options

The Data tab, shown in Figure 2-21, can be used to specify how SPSS handles a few special numeric situations:

✔ **Transformation and Merge Options:** Determines when — not how — results are calculated. You can have SPSS perform calculations immediately, or you can have it wait until it needs the number for something (either another calculation or a displayed value). Both methods have their advantages and disadvantages.

✔ **Display Format for New Numeric Variables:** Determines how many digits are used in the display of values and how many digits are to the right of the decimal. Width is the total number of characters, including the decimal point. The Decimal Places setting determines the number of digits that appear to the right of the decimal point. If the number of places to the right is too small, values are rounded to fit. If the number of places is too large, values are put into scientific notation.

✔ **Random Number Generator:** The method of implementing a random number generator on a computer has been a problem since someone discovered a need for such a number. Computers love to repeat themselves in a non-random way, so it is an interesting problem. SPSS lets you choose between two ways of doing it: the old way and the new way. If you have no interest in doing the same thing you did in older versions of SPSS (version 12 and earlier), use the Twister.

✔ **Set Century Range for 2-Digit Years:** A solution to the Y2K problem. I'll bet you thought that was all over, and it's true that the problem is gone. But the solutions are still with us and this is one of them. You put in two four-digit years here, and any two-digit value that you supply to identify a year is assumed to be between the two years you specify. This is mostly for old data. If you always use four digits for years in your data, this adjustment will never have to be made.

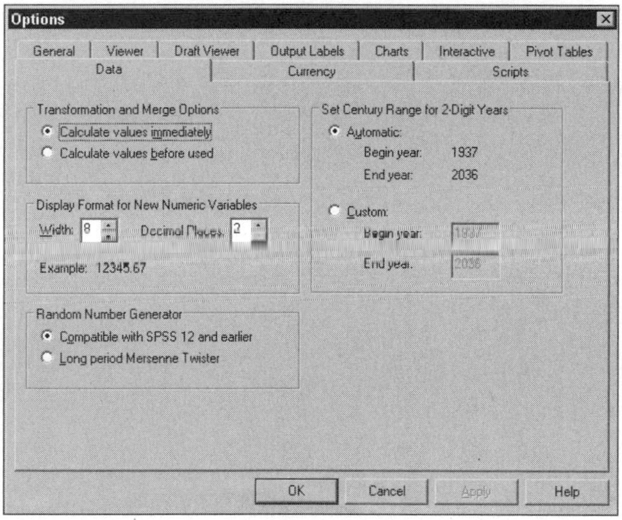

Figure 2-21: Data handling in SPSS can be varied within the limits of the settings of this tab.

Currency formats

Different parts of the world use different symbols and formats when writing about currency. The window shown in Figure 2-22 lets you specify the display format of your currency.

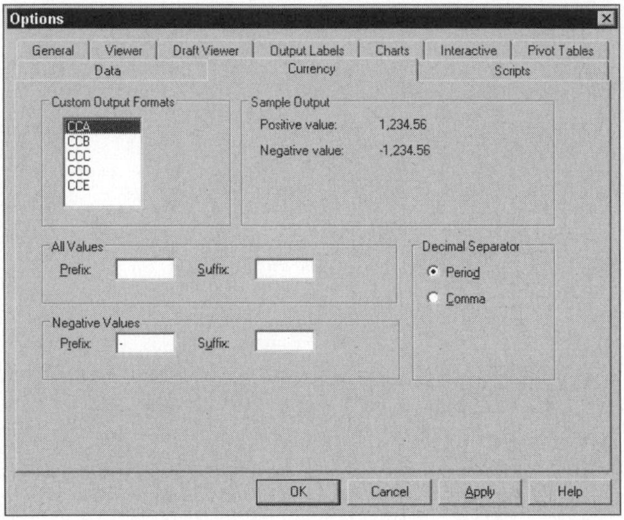

Figure 2-22:
Number
formats and
symbols can
be set so
SPSS
displays
things
correctly for
your
currency.

Following are the options in the Currency tab:

- ✔ **Custom Output Formats:** The default format for presenting currency values. The five formats have the unlikely names CCA, CCB, CCC, CCD, and CCE. Those are the only ones you can have, but that has to be enough for anybody. I mean, really, if you work with more kinds of money than that, buy another copy of SPSS. The calculations are always performed the same way — the differences are in the display. You can set the display configuration differently for each one and then switch among them as often as you like.

- ✔ **Sample Output:** Displays the printed format of positive and negative currency values. As you switch from one currency selection to another, and as you change the formatting of any of them, the sample displays examples of the format.

- ✔ **All Values:** Specifies characters that are displayed at the front and at the back of all values, such as a British pound sign and the cent mark.

- ✔ **Negative Values:** Specifies characters placed in the front and back of negative values. For example, some like to use < and > to surround negative money values.

- ✔ **Decimal Separator:** Many currency notations use commas instead of periods to denote the fractional portion of the amount.

Scripts options

Figure 2-23 displays the Scripts tab, which is used to determine some fundamental defaults about scripts.

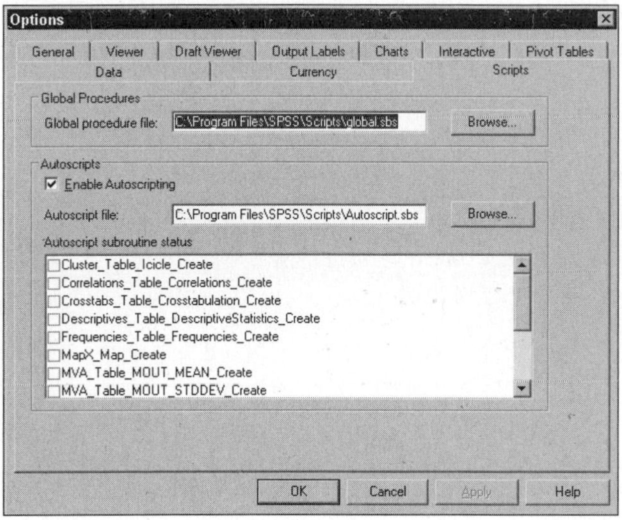

Figure 2-23:
This window configures global procedures and scripts that run automatically.

Don't mess with any of these until you've been writing scripts for a bit and know what you're doing.

- ✔ **Global Procedures:** The name of a file holding a library of procedures that can be called on by any scripts you write. These procedures are always present because you configure them as the default in this window. You can use the files that are provided or make up your own.

- ✔ **AutoScripts:** A script, if it is named correctly and stored in the file you name here, defines a global procedure. It runs automatically when you create an object of the associated type. To make it run, you have to select Enable AutoScripting, select one or more scripts from the list, and create an object of the correct type.

You shouldn't change any default option until you understand what that change will do. Such understanding will take a bit of experience using SPSS. You need to be aware that any change you make here will have an effect on all procedures you follow later.

Chapter 3

A Simple Statistical Analysis Example

In This Chapter

▶ Entering data into SPSS

▶ Performing an analysis

▶ Drawing a graph

This chapter goes through the process of entering some simple data into SPSS and then processing that data. I demonstrate various procedures for deriving results, using one subset of the data for some calculations and other parts of the data for other calculations. Finally, the results from these different calculations are displayed in different ways.

The data for this example is simple, as are the displays the data generates. The purpose of this chapter is not to present any great breakthrough in statistical analysis. Instead, I simply want to demonstrate the basic procedures you need to go through when operating SPSS.

When the Tanana at Nenana Thaws

This analysis is about an annual lottery that takes place in Alaska. Actually, it isn't called a lottery — it's called a classic, whatever that means.

I don't know whether the Tanana Classic is the oldest lottery in the United States (it began in 1917), but it's certainly the slowest. It has only one jackpot per year, and tickets for that jackpot are sold all across the state for months.

The lottery is simple enough. The citizens of the town of Nenana set up a large tripod on the ice in the middle of the Tanana River. From the top of the tripod, a tight line is stretched to a clock on a bridge. When the spring thaw comes, the tripod moves and the clock is triggered, stamping the exact minute. All the people who have selected the correct month, day, hour, and minute share the pot.

Many questions come to mind. What is the most likely date? What is the most likely time of day? Is there a trend? In the analysis that follows, we'll look at the answers to these questions and more.

By the way, the earliest the ice moved out was April 20 at 3:27 P.M. (in 1940), and the latest was May 20 at 11:41 A.M. (in 1964).

Entering the Data

SPSS can acquire its data from many sources. You can read data from a text file, a database, or a file produced by a program such as Access or Excel. This example does it the simplest way possible: The data is typed in using the editor window of SPSS. (I said simplest, not easiest.)

The data consists of dates and times. SPSS has a special date format that we'll be using later, but for now we'll enter the year, month, day, hour, and minute as separate numeric items. This keeps the example as simple as possible, and enables me to show you some different ways of manipulating numbers to reach conclusions.

The data definitions

The first job is to define the names, labels, and data types for the various fields of data, also known as the *variables*. Start the SPSS program. Choose Start⇨Programs⇨SPSS⇨SPSS 15.0 for Windows. An empty Data Editor window appears, as shown in Figure 3-1.

Figure 3-1:
The Data Editor window in Data View mode, before any data has been defined or entered.

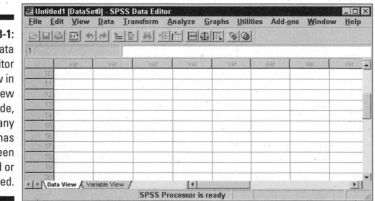

The layout shown in Figure 3-1 is the Data View mode, as indicated by the tab at the bottom of the window. We want to go to the other mode, so click the Variable View tab. The window now looks like the one in Figure 3-2.

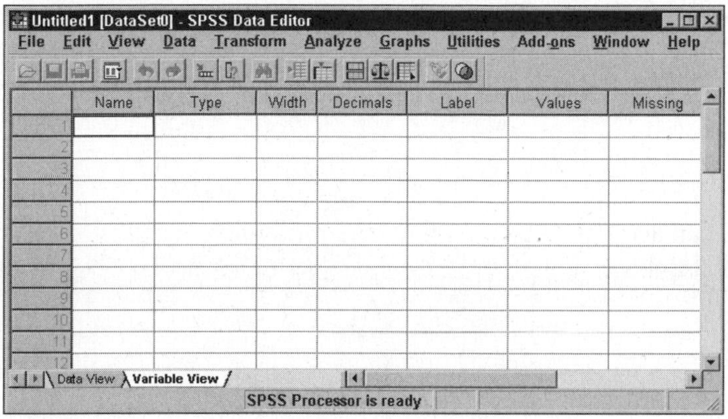

Figure 3-2: An empty Variable View window.

You use the Variable View window to define the names and types of variables, and you use the Data View window to enter the values for those variables. In Figure 3-3, I entered the variable definitions we'll be using in this example.

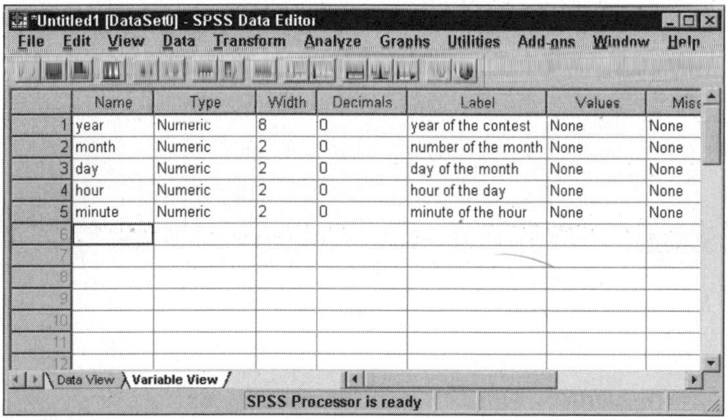

Figure 3-3: Definition of the variable names.

To enter the definitions, you type the name in the first column — the one labeled Name at the top — and then move the cursor down to the position for the next name in the list. You can move the cursor with either the keyboard arrow keys or by clicking the cell with the mouse.

When you move down to define a new variable, SPSS takes a wild guess at what you want in the cells you skipped, and automatically fills them in for you. Some of the guesses are right, and some are wrong. Stick with me here and I'll describe some of the fiddling around you'll need to do until your information matches that in Figure 3-3. For now, type the following entries in the Name column:

year

month

day

hour

minute

Every field has both a name and a label. One or the other is used as an identifying tag when data is displayed.

The name is usually shorter than the label. A short name is handy when you're displaying data in a tight format, such as a column heading or a bar chart label, and when you're writing equations in the two scripting languages supplied with SPSS. The label is intended to be more descriptive and can add clarity by being more descriptive in displays such as line graphs and pie charts.

In this example, all the fields are simple numerics, so SPSS guesses correctly about most of the attributes and fills them in for you. Most of the data you enter into SPSS will be numeric, although some numbers will be converted into names by SPSS. It's hard to calculate with things like "moonbeam" and "sure bubba," but I'll be showing you how to instruct SPSS to automatically change numbers into words and phrases later.

The width of most of the fields is 2 because they're two digits long. But, as an example, the year should be set to a length of at least 4 because we don't want to do Y2K all over again. Simply click the box (cell) for the year's Width column and type 4.

SPSS has set the number of digits to the right of the decimal point (the Decimals column) to 0 for all the numbers in our example, and that's what we want for this example. By the way, SPSS has a nifty date data type. I didn't use it here because I want to show you how to work with simple numbers. You find out about dates and some other special types and formats in Chapter 4.

When you type the label, you're not limited to the size of the cell that holds it. If you type a longer line, the box expands to take it all in. But don't write a thesis because you need something that will display nicely on your graphs and tables. Plus, you can always come back and change it. Type the following for the labels:

year of the contest

number of the month

day of the month

hour of the day

minute of the hour

Depending on how big you have your window, you may have to scroll to display columns to either side. To scroll, use the horizontal scroll bar at the bottom of the screen. I like to expand my window to the full screen, but that's probably because I'm easily distracted if I see other windows.

In the column labeled Missing, you specify whether or not values can be missing for this field. For example, if you are taking a survey on what color underwear people are wearing, you could assign a number to each color, but you are bound to come across someone who isn't wearing any, so you'll need to define the field in such a way that it allows for such items to be missing. By default, SPSS does not allow for missing data, so the default is None.

The default column width for a data item is 8, and that's okay for this example. You can make the columns smaller, if you prefer, but you need to make sure the columns are big enough to hold your largest data item or its name. This is the amount of space that's going to be allocated when SPSS constructs charts and tables. If you set the size too small, the data or the variable name will be cut for some displays.

The alignment specifies whether the data should be aligned on the right, shoved over to the left, or placed in the center. Choose whatever you like. This is determined by personal preference, a lousy sense of design, and bad taste.

The last column on the right is labeled Measure. It can be set to Scale, which is the default, Ordinal, or Nominal. Leave it set to Scale. Scale is an amount or size — it's just a regular number — and works fine for what we're going to do. Ordinal has to do with things that have a specific order. Nominal values are used to tag things as belonging to categories.

The actual data

Click the Data View tab, which is at the bottom of the Data Editor window, and the window changes to look like the one shown in Figure 3-4. The label names you entered in the Variable View window appear at the top of the columns. This window is now ready for you to enter numeric data.

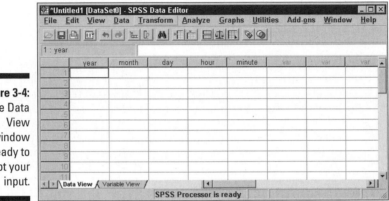

Figure 3-4:
The Data
View
window
ready to
accept your
input.

In Figure 3-4, notice the numbers down the left side of the window. This is the SPSS way of numbering rows, which are also called *cases*. If you use the scroll bar on the right side of the window to scroll down, you'll see these numbers change. You can think of these numbers as a roadmap to the layout in the window so you can keep track of where you are.

However, don't trust the numbers to identify your data. If you move your data from place to place in the grid, the numbers on the left don't move with it. That means that if you insert a row, delete a row, or simply sort your data in a different way, the numbers on the left will associate with different sets of values. Your case numbers will all be different. If you need to identify a case in a manner that does not change when the organization of the cases changes, you must add a field for identity and enter your own identifying numbers. You can see exactly how to do this in Chapter 7.

All the values that need to be entered for this example are in Table 3-1. However, you can be lazy if you want to because I've already entered all the numbers. All you have to do is load the file that holds them by choosing File➪Open➪Data and selecting nenana.sav. But even if you decide to read them in from the file, you should at least start out by entering a few so you can see how SPSS data entry works. I talk about loading the file a little later.

Table 3-1   The Dates and Times of the Nenana Thaw

4/20/1940, 15:27	4/29/2003, 18:22	5/3/1919, 14:33	5/6/1974, 15:44	5/10/1975, 13:49
4/20/1998, 16:54	4/30/1917, 11:30	5/3/1941, 13:50	5/6/1977, 12:46	5/10/1982, 17:36
4/23/1993, 13:01	4/30/1934, 14:07	5/3/1947, 17:53	5/7/1925, 18:32	5/11/1918, 09:33
4/24/1990, 17:19	4/30/1936, 12:58	5/4/1944, 14:08	5/7/1965, 19:01	5/11/1920, 10:46
4/24/2004, 14:16	4/30/1942, 13:28	5/4/1967, 11:55	5/7/2002, 20:27	5/11/1921, 06:42
4/26/1926, 16:03	4/30/1951, 17:54	5/4/1970, 10:37	5/8/1930, 19:03	5/11/1924, 15:10
4/26/1995, 13:22	4/30/1978, 15:18	5/4/1973, 11:59	5/8/1933, 19:30	5/11/1985, 14:36
4/27/1988, 09:15	4/30/1979, 18:16	5/5/1929, 15:41	5/8/1959, 11:26	5/12/1922, 13:20
4/28/1943, 19:22	4/30/1981, 18:44	5/5/1946, 16:40	5/8/1966, 12:11	5/12/1937, 20:04
4/28/1969, 12:28	4/30/1997, 10:28	5/5/1957, 09:30	5/8/1968, 21:26	5/12/1952, 17:04
4/28/2005, 12:01	5/1/1932, 10:15	5/5/1961, 11:30	5/8/1971, 21:31	5/12/1962, 21:23
4/29/1939, 13:26	5/1/1956, 23:24	5/5/1963, 18:25	5/8/1986, 22:50	5/13/1927, 05:42
4/29/1953, 15:54	5/1/1989, 20:14	5/5/1987, 15:11	5/8/2001, 13:00	5/13/1948, 11:13
4/29/1958, 14:56	5/1/1991, 12:04	5/5/1996, 12:32	5/9/1923, 14:00	5/14/1949, 23:39
4/29/1980, 13:16	5/2/2000, 10:47	5/6/1928, 16:25	5/9/1955, 14:13	5/14/1992, 06:26
4/29/1983, 18:37	5/2/1960, 19:12	5/6/1938, 20:14	5/9/1984, 15:33	5/15/1935, 13:32
4/29/1994, 23:01	5/2/1976, 10:51	5/6/1950, 16:14	5/10/1931, 09:23	5/16/1945, 09:41
4/29/1999, 21:47	5/2/2006, 17:29	5/6/1954, 18:01	5/10/1972, 11:56	5/20/1964, 11:41

You should now be displaying the Data View window. To enter a number, simply click a position with the mouse and then type the number that you want to go into that square.

When I entered the data, I duplicated a row that was already there and then made changes to it. This was handy because the month and day of the new entry were often the same as the duplicated entry. To duplicate a row, select the row you want to copy by clicking the number at the left of the row. One click selects the entire row. Then choose Edit➪Copy. Next, select the row where you want the data to go and choose Edit➪Paste. If your target row already contains data, the new data overwrites it.

Suppose you want to insert a new row of data in front of some you already have. First, select the row that is in the place you want the new one to go and choose Edit➪Insert Cases. This opens a blank row. Then you can either copy or type new data into the blank row.

When you're finished, you can scroll up and down and see different parts of the data, as shown in Figure 3-5.

	year	month	day	hour	minute	var	var	var
70	1984	5	9	15	33			
71	1931	5	10	9	23			
72	1972	5	10	11	56			
73	1975	5	10	13	49			
74	1982	5	10	17	36			
75	1918	5	11	9	33			
76	1920	5	11	10	46			
77	1921	5	11	6	42			
78	1924	5	11	15	10			
79	1985	5	11	14	36			
80	1922	5	12	13	20			

Figure 3-5:
The data freshly entered into SPSS.

When you're entering your own data, select a file name early in the process and choose File➪Save to write everything to the file from time to time. If you don't do this, a simple computer crash could lose all your data. That sort of thing is not good for your blood pressure.

By the way, if you've scrolled all the way down, you've noticed that there is a bottom to the list of numbered rows. Don't worry about it. As you enter data, the bottom extends so you never hit a limit.

If you've elected not to enter the data by hand, and instead want to load it from the file, choose File➪Open➪Data, then navigate to wherever you stored the `nenana.sav` file, as shown in Figure 3-6. Depending on how your Windows system is configured, the name may be chopped off in your display and appear only as `nenana`. It's not abnormal for Windows to change file names this way. The book's Introduction tells you how and where you can get the files.

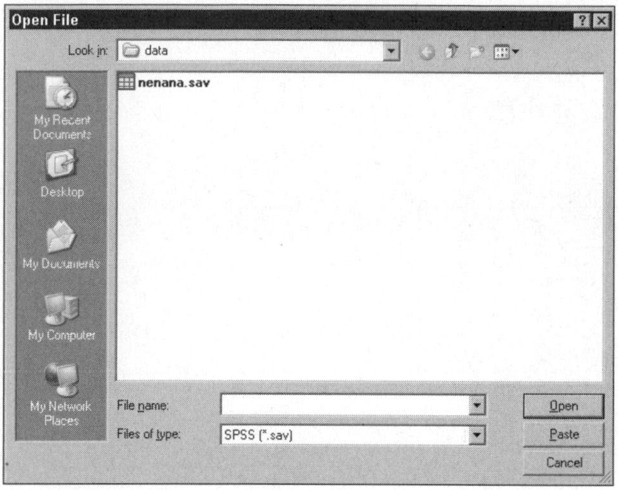

Figure 3-6:
Loading
an SPSS
data file.

The Most Likely Hour

Now that we have the data in SPSS, let's do something simple. The following procedure finds the mean of the hours in an attempt to determine the hour of the day when the ice is most likely to melt. It makes sense that this would probably be in the daytime because the sun is warming both the air above the ice and the flowing water below the ice.

To find the most likely hour (ignoring the minutes for now), perform the following steps:

1. **Choose Analyze➪Descriptive Statistics➪Descriptives.**

2. **In the box on the left, select** hour of the day **(one of your variable labels) and then click the little button in the middle of the window.**

   The label moves to the right, as shown in Figure 3-7.

3. **Click the Options button.**

4. **Select the Mean, Std deviation, Minimum, and Maximum check boxes, as shown in Figure 3-8.**

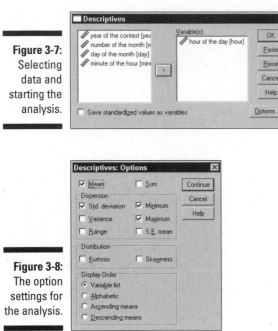

Figure 3-7:
Selecting
data and
starting the
analysis.

Figure 3-8:
The option
settings for
the analysis.

5. **Click Continue.**

6. **Click the OK button in the upper-right corner of the window in Figure 3-7.**

   SPSS Viewer appears and displays information about the analysis, including the results. A detailed description of all this information is in Chapter 8. For now, use the scroll bars to locate the result in the box at the bottom of the window, as shown in Figure 3-9. The mean (not the average, but nearly the same thing) shows the hour as 14.60, which is in the afternoon. That makes sense, because that's near the warmest part of the spring day.

Figure 3-9:
The results
of the
simple hour
analysis.

Descriptive Statistics

	N	Minimum	Maximum	Mean	Std. Deviation	Variance
hour of the day	90	5	23	14.60	4.069	16.557
Valid N (listwise)	90					

Inside the box, the text on the far left is the label you gave to the variable. The column labeled N is the number of data items included in the calculations. You can tell from the minimum and maximum that the earliest the ice has ever let go was during the 5 o'clock hour in the morning, but it has also been known to happen after 11 at night.

The values for the standard deviation and variance are calculated according to their variation from a perfect fit on a bell curve. The two values are different ways of looking at the same thing — the standard deviation squared results in the variance.

There is more bell curve stuff to diddle with. Go back through the same procedure again, but this time change the options in Step 4 to include Kurtosis and Skewness. Those are not rude words and, no, I didn't just make them up. They are part of statistics. As shown in Figure 3-10, the results have two new values.

Figure 3-10:
New
analysis
showing
kurtosis and
skewness.

Descriptive Statistics

Mean	Std.	Variance	Skewness		Kurtosis	
Statistic	Statistic	Statistic	Statistic	Std. Error	Statistic	Std. Error
14.60	4.069	16.557	.086	.254	-.480	.503

Both values also have to do with the bell curve. *Skewness* represents the symmetry of the data. A positive skewness indicates that more of the data appears to the high end, or the right, on the graph. A negative value indicates a skew to the lower values. *Kurtosis* has to do with the flatness of the curve. If the data implies a curve flatter than the bell curve, the kurtosis value is negative. If, on the other hand, the data inscribes a curve that is more pointed on top than the bell curve, the kurtosis value is positive.

Transforming Data

In the previous example, we looked at only the hours, but it's possible to also include the minutes. Clock arithmetic is tricky (it's that 60 minutes per hour thing), but SPSS can work with it if you tell it what you're doing.

In the next example, we'll combine the separate hour and minutes fields into a new field that contains both. SPSS is good at transforming data this way. To build the new field, do the following:

1. **In the SPSS Data Editor window, choose Transform⇨Date and Time Wizard.**

   The window shown in Figure 3-11 appears.

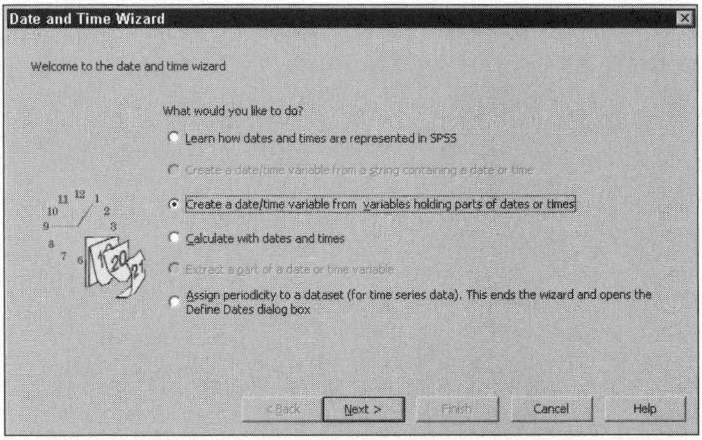

Figure 3-11:
The Date
and Time
Wizard.

2. **Select the option titled Create a Date/Time Variable from Variables Holding Parts of Dates or Times.**

3. **Click Next.**

4. **Put the names of the variables into the appropriate fields.**

   We want only the hours and minutes, so ignore the others. You move them by selecting the one you want from the list on the left and then clicking the triangle next to the place you want it to go. When you're finished, the screen should look like Figure 3-12.

5. **Click Next.**

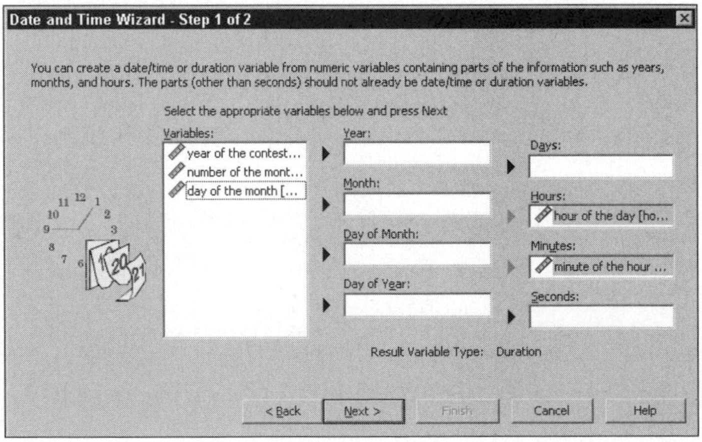

Figure 3-12:
Selection
of the
variables
from which
time is
structured.

6. **Enter a name and a label for the variable. Also select a display format from the list.**

   To follow along with the example, type **time** in the Result Variable box, type **hour and minute** in the Variable Label box, and select **hh:mm** in the Output Format list, as shown in Figure 3-13.

7. **Select the Create the Variable Now option, and then click the Finish button.**

   You've created your new time data field. The result is shown in Figure 3-14.

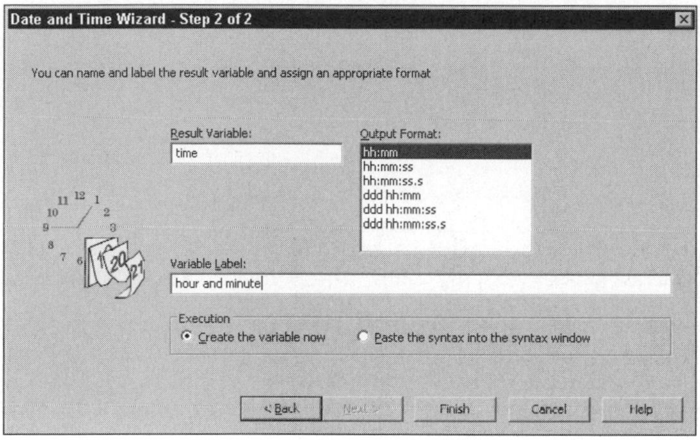

Figure 3-13: The name and display format for times.

Now follow the same procedure as before by choosing Analyze⇨Descriptive Statistics⇨Descriptives. But in Step 4, select only the hours so you can see how SPSS handles different combinations of values. In the results, look at the difference in the two means. When the minutes are included, the mean moved to a time a bit later (as one would expect). It is now at 3:03 PM. Whether the difference is statistically significant is up to you.

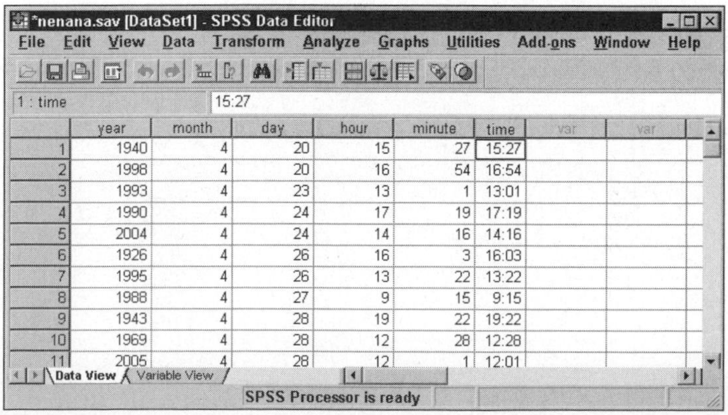

	year	month	day	hour	minute	time	var	var
1	1940	4	20	15	27	15:27		
2	1998	4	20	16	54	16:54		
3	1993	4	23	13	1	13:01		
4	1990	4	24	17	19	17:19		
5	2004	4	24	14	16	14:16		
6	1926	4	26	16	3	16:03		
7	1995	4	26	13	22	13:22		
8	1988	4	27	9	15	9:15		
9	1943	4	28	19	22	19:22		
10	1969	4	28	12	28	12:28		
11	2005	4	28	12	1	12:01		

Figure 3-14: The Data Editor window with the new time field.

The Two Kinds of Numbers

With this example data so far, we have dealt with continuous variables. *Continuous variables* are amounts and distances, such as age, gallons of gas, and the number of beans in a jar. The other type of number is *categorical values.* Here you will find things such as yes and no (where, for example, yes is 1 and no is 0) and types of balls (where 1 is a football, 2 is a soccer ball, 3 is a snooker ball, and so on). Each value represents a category.

All the variables in this example, except the number indicating the month, are continuous variables. We tend to think of the months by their names instead of numbers. You must use the number of the month to do any calculations, but if you want the name to be displayed, you have to assign a descriptive name for each possible value. That's easy to do in this case because we have only two values: 4 and 5.

To add identifiers for the values, do the following:

1. **In the Data Editor window, click the Variable View tab and then select the cell in the Values column of the variable holding the month values.**

2. **Click the button that appears in the cell.**

   The dialog box shown in Figure 3-15 appears.

3. **For each value, enter the value and the name you want associated with it, and then click Add.**

   The value, with its identifier, appears in the list, as shown in Figure 3-16.

4. **After you've added all the values you want to define, click OK.**

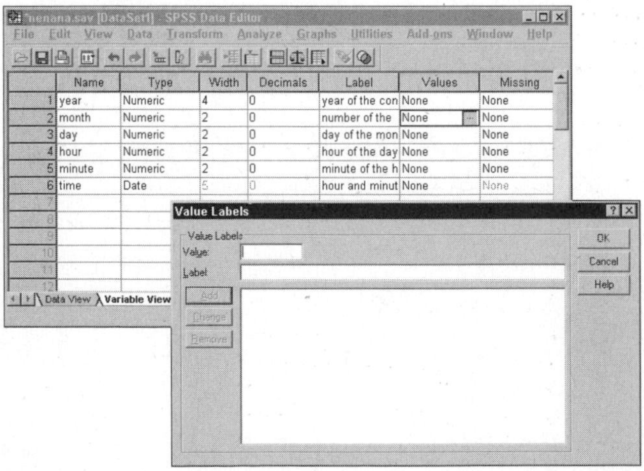

Figure 3-15: This is where you define descriptive tags for values in variables.

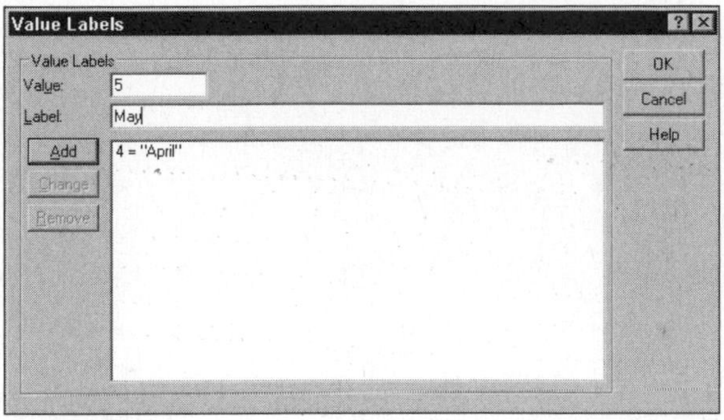

Figure 3-16:
One name
has been
added for a
value and
another one
is being
entered.

If you look at the screen, it appears that nothing has changed. Do you have a feeling that you did all that work with no result? That's not so. The result will show up in your output and help you make a lot more sense of your results. For example:

1. **Choose Graphs⇨Legacy Dialogs⇨Pie.**

   The window shown in Figure 3-17 appears.

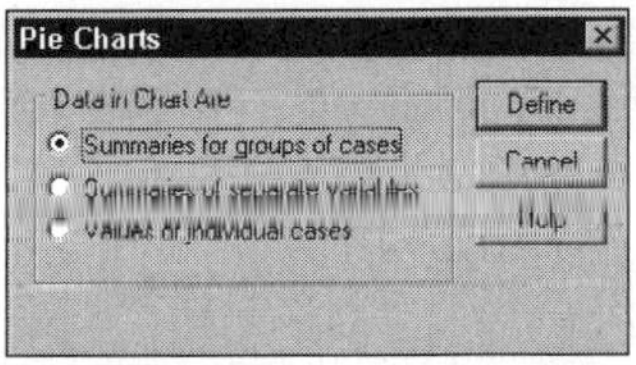

Figure 3-17:
Select the
type of
data to be
displayed
in the pie
chart.

2. **Select the Summaries for Groups of Cases option, and then click the Define button.**

   The window shown in Figure 3-18 appears.

3. **In the column on the left, select** number of the month, **then click the triangle to the left of Define Slices By, as shown in Figure 3-18.**

4. **Click the OK button.**

   SPSS Viewer appears, as shown in Figure 3-19.

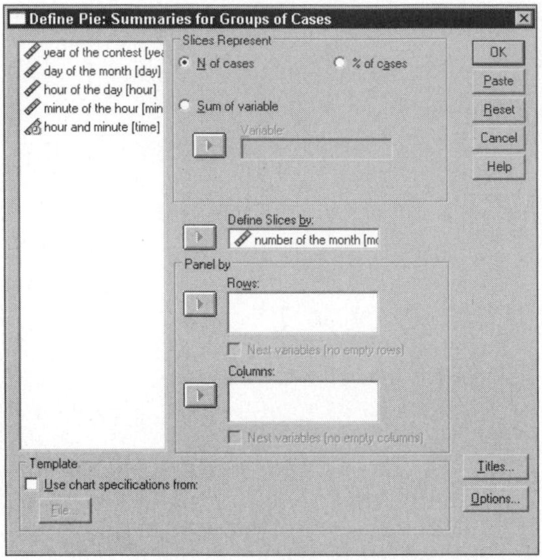

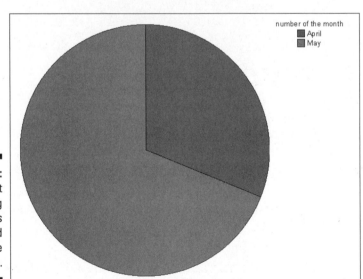

Figure 3-18:
You can
select the
variables
you want
for the pie
divisions.

Figure 3-19:
A pie chart
including
the names
you defined
for the
values.

The Day It Is Most Likely to Happen

You already know that the ice is most likely to move in the warmer part of the day. A quick graph can show you whether or not there's a most likely day as well. To get a quick bar graph, do the following:

1. **Choose Graphs⇨Legacy Dialogs⇨Bar.**

   The dialog box shown in Figure 3-20 appears.

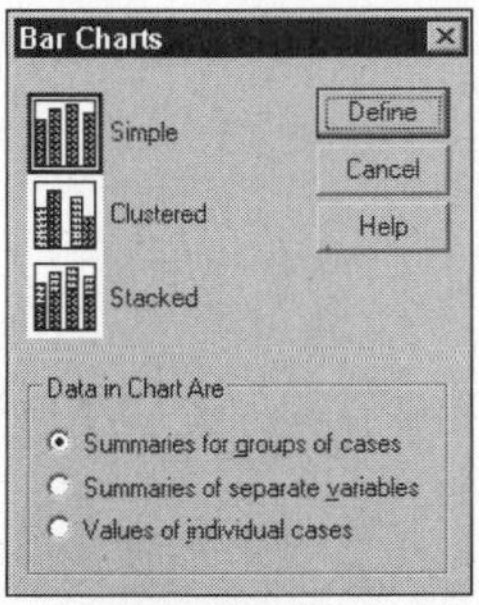

Figure 3-20:
You can
select the
fundamen-
tals of the
bar chart
you want.

2. **Select the Simple bar chart and the Summaries for Groups of Cases option, and then click the Define button.**

3. **For Bars Represent, select N of cases, which means the bars will represent the number of cases. Also set the Category Axis to be the day of the month (day) and set the Rows to be the number of the month (month), as shown in Figure 3-21.**

   The exact meanings of these terms and settings are explained in Part III, which covers graphs.

4. **Click the OK button, and the bar chart in Figure 3-22 appears.**

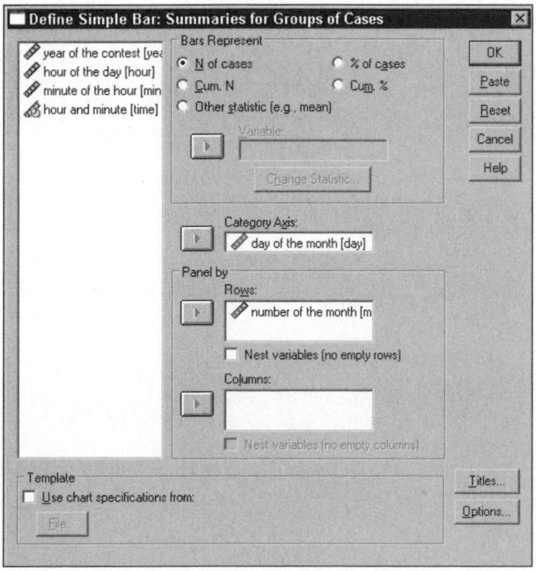

Figure 3-21:
Selecting
the data to
include in
the bar
chart.

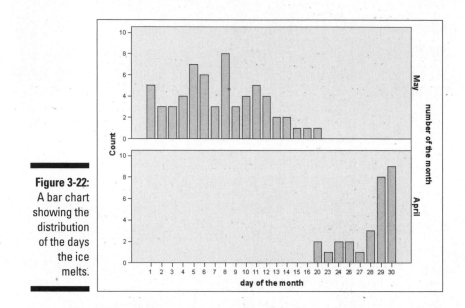

Figure 3-22:
A bar chart
showing the
distribution
of the days
the ice
melts.

The resulting chart shows which days in the past were most often the ones
on which the ice moved. There is no obvious trend that I can see. However,
you might want to experiment with different analysis displays and try to find
a pattern.

Part II
Getting Data into and out of SPSS

The 5th Wave   By Rich Tennant

"Please answer the following survey questions about our company's performance with either, 'Excellent', 'Good', 'Fair', or 'I'm Really Incapable of Appreciating Someone Else's Hard Work.'"

In this part . . .

The purpose of SPSS is to crunch numbers to come up with other numbers. To do the crunching, you have to get the numbers into the program. After the crunching is finished, you have to get the numbers out so you can see them. In fact, with the single exception of robotics, the sole purpose of every program in the world is to contain numbers and display them to a human.

Input can be tedious, but SPSS has ways of helping ease the pain. Regular output is automatic, but you can do special things for irregular output.

Chapter 4

Entering Data from the Keyboard

• •

• •

To process your data, you have to get it into the computer. Entering data has been a problem with computers since the beginning. No matter how you decide to get your numbers into SPSS, at some point someone has to type them (unless they come from the automatic monitoring of a machine). SPSS can read data from other places. You can also type directly into SPSS and, if you want, copy to places other than SPSS later.

Entering data into SPSS is a two-step process. First, you define what sort of data you will be entering, then you enter the actual numbers. After you see how data entry works in SPSS, you'll realize you have some pretty nifty windows to help you.

You organize your data into cases in which each case is made up of a collection of variables. First, you define the characteristics of the variables that make up a case, and then you enter the data into the variables to make up the contents of the records.

The Variable View Is for Entering Variable Definitions

You use the Variable View, shown in Figure 4-1, to define the names and characteristics of variables. This is where you always start if you plan on entering data into SPSS. You get to this window by clicking the Variable View tab at the bottom of the Data Editor window of SPSS. As you can see in Figure 4-1, every characteristic you can define about your variables is named at the top of the window. All you have to do is enter something in each column for each variable.

Figure 4-1:
You use
Variable
View to
define the
character-
istics of
variables.

The Variable View window is just for describing the variables. The entry of the actual numbers comes later.

Each variable characteristic has a default. So if you don't specify a characteristic, SPSS will fill in one for you. But what it selects may not be what you want, so let's look at all the possibilities.

Name

The cell on the far left is where you enter the name of the variable. This is the short descriptor such as age, income, sex, or odor. A longer descriptor, called a *label,* comes later. You could type longer names here, but you should keep them short because they will be used in named lists and as identifier tags on the data graphs and such where the format can be a bit crowded. Names that are too long can cause the output from SPSS to be garbled or truncated.

If you lose your head and assign a name that turns out to be too long, or if you misspell it or something, you can always pop back into Variable View and change it. One of the nice things about SPSS is that it allows you to quickly correct mistakes. I like that. I had to hide a lot of them for the screen shots in this book.

The SPSS default names are never descriptive. Believe me, any name you make up is better.

You can use some bizarre characters in a name, such as @, #, and $. You can also use the underscore character (_) and numbers, but you need to start every name with an uppercase or lowercase letter and you can't include blanks anywhere in a name. If you decide you want to use some screwy characters in a name, go ahead and try it. SPSS will threaten you with legal action but it never does anything about it other than make you type something else.

If you might be exporting data to another application, make sure the names you use are in a form acceptable to that application. Watch out for special characters.

Type

Most of the numbers you enter will be just regular numbers. Some, however, will be a special type, such as currency, and some will be displayed in a special format. Others, such as dates, require special procedures for calculation. You simply specify what type you have and SPSS takes care of the details for you.

Click the cell in the Type column you want to fill in, and a button with three dots appears on its right. Click that button and the dialog box shown in Figure 4-2 appears.

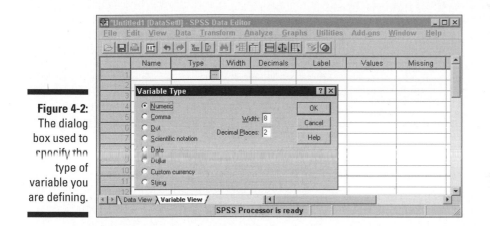

Figure 4-2: The dialog box used to specify the type of variable you are defining.

You can choose from the following predefined types of variables:

- **Numeric:** Standard numbers in any recognizable form. The values are entered and displayed in the standard form, with or without decimal points. Some values could be formatted in standard scientific notation, with an embedded *E* to represent the start of the exponent. The Width is the total number of all characters in a number, including any positive and negative signs, and the exponent indictor. Decimal Places specifies the number of digits displayed to the right of the decimal point, not including the exponent.

- **Comma:** Numeric values with commas inserted between each group of three digits. The format includes a period as a decimal point. The Width is the total width of the number, including all commas and the decimal point. Decimal Places specifies the number of digits to the right of the decimal point. You may enter data without the commas, but SPSS will insert them when it displays the value. Commas are never placed to the right of the decimal point.

✔ **Dot:** Same as Comma, except a period character is used to group the digits into threes, and a comma is used for the decimal point.

✔ **Scientific Notation:** A numeric variable that always includes the *E* to designate the power of ten exponent. The base, the part of the number to the left of the *E,* may or may not contain a decimal point. The exponent, the part of the number to the right of the *E,* which may or may not also contain a decimal, is an exponent of the number 10, which is then multiplied by the base to produce the actual number. You may enter *D* or *E* to mark the exponent, but SPSS always displays the number using *E.* For example, the number 5,286 can be written as 5.286E3. To represent a small number, the exponent can be negative. For example, the number 0.0005 can be written as 5E-4. This format is useful for very large or very small numbers.

✔ **Date:** A variable that can include the year, month, day, hour, minute, and second. When you select Date, the dialog box shown in Figure 4-3 appears. In the list on the left, choose the format that best fits your data. Your selection determines how SPSS will format the contents of the variable for display. This format also determines, to some extent, the form in which you enter the data. You can enter the data using slashes, colons, spaces, or other characters. The rules are loose — if SPSS doesn't understand what you enter, it tells you and you can reenter it another way. If you select a format with a two-digit year, SPSS accepts and displays the year that way, but it will use four digits to perform calculations. The first two digits (the number of the century) will be selected according to the configuration you set by choosing Edit➪Options and then clicking on the Data tab.

✔ **Dollar:** Dollar values are always displayed with a leading dollar sign and a period for a decimal point, and may include commas to collect the digits in groups of threes. You select the format and its Width, as shown in Figure 4-4. The format choices are similar, but it is important that you choose one that is compatible with your other dollar variable definitions so they line up when you print and display monetary values in output tables. The Width and Decimal Places settings help with vertical alignment in the output, no matter how many digits you include in the format itself. No matter what format you choose, you can enter the values without the dollar sign and the commas; SPSS will insert those for you.

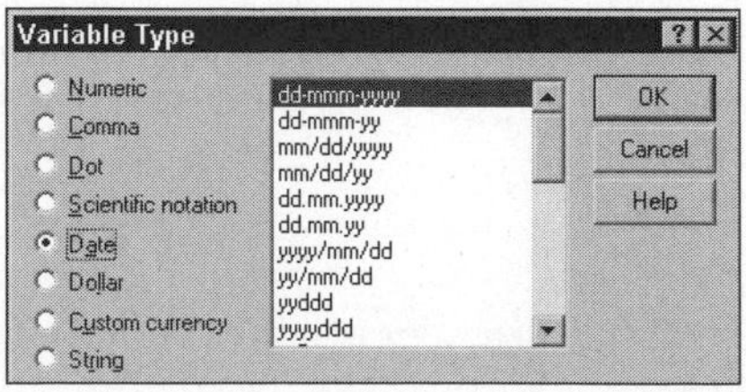

Figure 4-3:
Selecting a
date format
also selects
which items
are included.

✔ **Custom Currency:** The five custom formats for currency are named CCA, CCB, CCC, CCD, and CCE, as shown in Figure 4-5. The details of these formats can be viewed and modified by choosing Edit⇨Options and then clicking the Currency tab. Fortunately, you can modify the definitions of these custom formats as often as you like without fear of damaging your data. As with the Dollar format, the Width and Decimal Places settings are primarily for aligning the data when printing a report.

✔ **String:** A freeform non-numeric item. Because it is non-numeric, the contents of a variable of this type can never be used for calculations. You can specify any number of any characters up to the maximum length you specify, as shown in Figure 4-6. A variable of this type could be used as a descriptor or an identifier of a particular case.

Figure 4-4:
The different dollar formats mostly specify the number of digits to be included.

Figure 4-5:
Five custom currency formats are available.

Figure 4-6:
A freeform type never used in calculations.

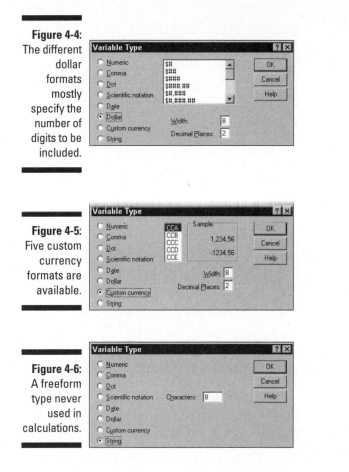

Width

The width setting in the definition of a variable determines the number of characters used to display the value. If the value to be displayed is not large enough to fill the space, the output will be padded with blanks. If it is larger than you specify, it will be reformatted to fit or asterisks will be displayed.

Certain type definitions allow you to set a width value. The width value you enter as the Width definition is the same as the one you entered when you define the type. If you make a change to the value in one place, SPSS changes the value in the other place. The two are the same.

You can do one of three things

- ✔ Skip this cell and accept the default (or the number you entered previously).
- ✔ Enter a number and move on.
- ✔ Use the up and down arrows that appear in the cell to select a numeric value.

Decimals

The number of decimals is the number of digits that will appear to the right of the decimal point when the value is displayed. This is the same number of decimal digits that you may have specified when you defined the variable type. If you entered a number there, it will appear here as the default. If you enter a number here, it will change the one you entered for the type. They are the same.

You can do one of three things

- ✔ Skip this cell and accept the default (or the number you entered earlier).
- ✔ Enter a number and move on.
- ✔ Use the up and down arrows that appear in the cell to select a numeric value.

Label

The name and the label serve the same basic purpose: They are descriptors that identify the variable. The difference is that the name is the short identifier and the label is the long one. You need one of each because some output formats work fine with a long identifier and other formats need the short form.

You can use just about anything for the label. What you choose has to do with how you will be using your data and what you want your output to look like. For example, the name may be *sex* and the longer label may be *Boys and Girls, Men and Women,* or simply *Gender.*

The length of the label is not determined by some sort of software requirement. However, output looks better if you use short names and somewhat longer labels. Each one should make sense standing alone. After you produce some output, you may find that your label is lousy. That's okay. It's easy to change. Just pop back to the Variable View and make the change. The next time you produce output, the new label will be used.

You can also just skip defining a label. If you don't have a label defined for a variable, SPSS will use the name for everything.

Value

The Values column is where you assign labels to all the possible values of a variable. If you select a cell in the Values column, a button with three dots appears. Clicking that button displays the dialog box shown in Figure 4-7.

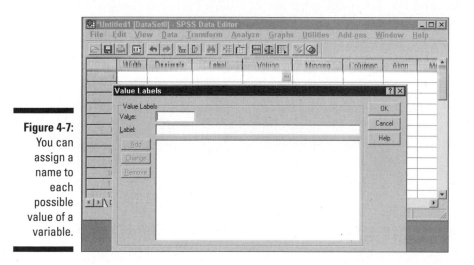

Figure 4-7: You can assign a name to each possible value of a variable.

Normally, you make one entry for each possible value that a variable can assume. For example, you could have 1 for Male and 2 for Female. Or you could have 0 for No, 1 for Yes, and 2 for Undecided. When you enter definitions for values, the values appear as strings of characters instead of simple numbers in your output displays.

To define a label for a value:

1. **In the Value box, enter the value.**

2. **In the Label box, enter a label.**

3. **Click the Add button.**

   The value and label appear in the large text block. To change or remove a definition, simply select it in the text box and make your changes.

4. **Repeat Steps 1–3 as needed.**

5. **Click the OK button to save the value labels and close the window.**

You can always come back and change the definitions using the same process you used to enter them. The window will reappear filled in with all the definitions and you can change or modify the list.

Missing

You can specify whether a value can be missing for a variable in a case. That is, you may have values for the other variables in any given case, but nothing for this one. Click a cell in the Missing column, and the dialog box shown in Figure 4-8 appears.

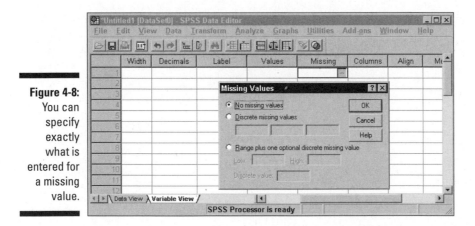

Figure 4-8:
You can specify exactly what is entered for a missing value.

You can specify more than one missing value. You can specify what number will be entered to indicate that the value is missing. You can have more than one value to indicate a missing value because you may want to indicate the reason why the value is missing. For example, if you are conducting a survey on dirigible ownership, one person might say they forget, another might say

"None of your business," and another might say "37," but you know that no one has 37 dirigibles.

If you specify that a value is representing a missing value, that value is not included in general calculations. However, during your analysis, you will be able to determine how many values are missing for each of your reasons. You can specify up to three specific values (called *discrete values*) to represent missing data, or you can specify a range of numbers along with one discrete value, all to be considered missing. The only reason you would need to specify a range of values is if you have lots of reasons why data is missing and want to track them all.

Columns

Columns is where you specify the width of the column you will use to enter the data. The folks at SPSS could have used the word *Width* to describe it, but they already used that term for the width of the data itself. A better name might have been the two words *Column Width,* but that would have been too long to display nicely in this window, so they just called it *Columns.* To specify the number of columns, select a cell and enter the number.

Align

The Align column determines the position of the data in its allocated space. The data can be left aligned, right aligned, or centered. You've defined the width of the data and the size of the column in which the data will be displayed; the alignment determines what is done with any space left over.

When you select a cell in the Align column, a list appears and you can choose one of the three alignment possibilities, as shown in Figure 4-9. Aligning to the left means inserting all blanks on the right, aligning right inserts all the extra spaces on the left, and centering the data splits the spaces evenly on each side — but I don't know what it does if an odd space is left over. I also worry about things like the number of seeds in a tomato and where the clouds go at night.

Measure

Your value will specify the measure of something in one of three ways. When you click a cell in the Measure column, you can select one of the choices in the list shown in Figure 4-10.

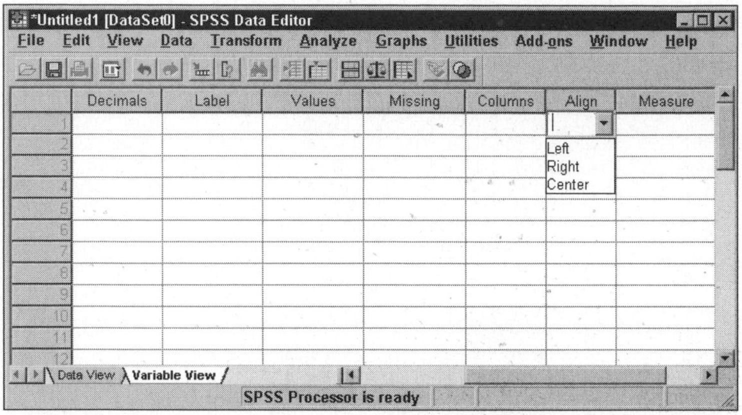

Figure 4-9:
Values can
be justified
right or
left or
positioned
in the
center.

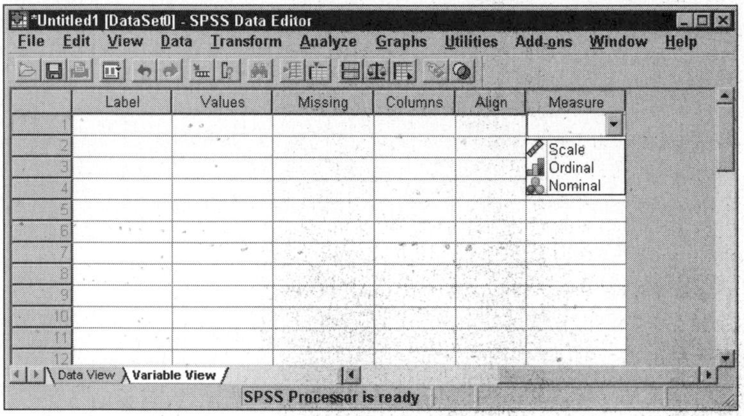

Figure 4-10:
The type of
measure-
ment being
made by
the values
in this
variable.

You have three choices:

- **Scale:** A number that specifies a magnitude. It can be distance, weight, age, or a count of something. Most numbers fall in this category. The technical name for this type of number is cardinal, but SPSS uses Scale to keep life simple.

- **Ordinal:** These numbers deal with the position (order) of something in a list. For example, first, second, and third are ordinal numbers.

- **Nominal:** Numbers that specify categories or types of things. You can have 0 represent Disapprove and 1 represent Approve. Or you can use 1 to mean Fast and 2 to mean Slow.

The Data View Is for Entering and Viewing Data Items

After you've defined all the variables you need for each case, switch the display to the Data View so you can begin typing the data. You make the switch by clicking the Data View tab at the bottom of the window. When you do, the Data Editor window appears.

At the top of the columns in Figure 4-11, you can see some names I chose for variables. By switching to Data View, the window became ready to receive entered data and verify that what is entered matches the specified format and type of the data.

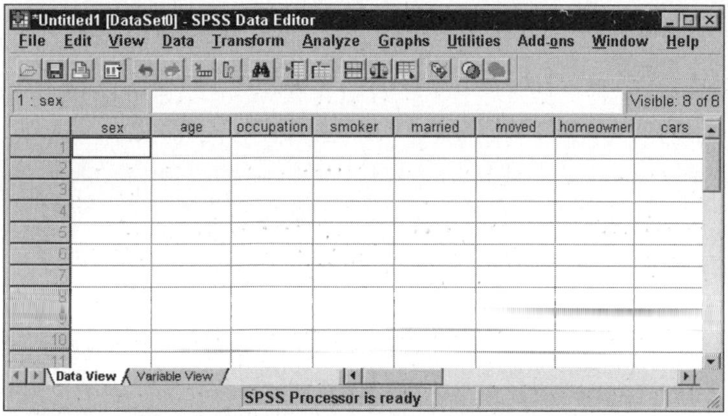

Figure 4-11: The Data Editor window ready to accept new data.

Entering data into one of these cells is straightforward: You simply click the cell and start typing.

If something is already in a cell and you want to change it instead of just type over it, look up toward the top of the window, where you'll see the name of the variable and the currently selected value. Click the value in the field at the top and you can edit it right there. You can do all the normal mouse and keyboard stuff there too — you can use the Backspace key to erase characters, or select the entire value and type right over it.

If you are a lousy or inexperienced mouse driver, take some time to experiment and figure out how to edit data. Lots of software use these same editing techniques, so becoming proficient now will pay you dividends later.

If your data is already in a file, you might be able to avoid typing it in again by reading that file directly into SPSS. For more information, see Chapter 5.

Don't take chances. As soon as you type a few values, save your data to a file by choosing File⇨Save As. Then choose File⇨Save throughout the process of entering data, and you won't be ruined when the computer crashes unexpectedly.

We all have to go back and refine our variable definitions from time to time. That's normal. When you come across something that doesn't do what you want it to, just switch back to Variable View and correct it. Nobody but you and SPSS will ever know about it, and SPSS never talks.

Filling In Missed Categorical Values

Now that you have defined your variables and entered your data, you might want to check that you have names defined for all ordinal and nominal values, and that you have defined the correct measures for them. SPSS can help by scanning your data, finding values for which you don't have definitions, and pointing those out in a friendly way.

The following steps use an existing file to walk through a demonstration:

1. **Choose File⇨Open⇨Data to load the file named `Cars.sav`.**

   This file came with your installation of SPSS and is found, along with a number of other files, in the same directory in which you installed SPSS. You can load any of these data files, but `Cars.sav` is the one used in this demonstration. If you load this file while you already have some other data showing in the window, SPSS will open a new Data Editor window to display the new information, but your existing data will not be lost.

   When you open this data file — or any data file, for that matter — SPSS will open an SPSS Viewer window to tell you that it has opened a file (or the information could be displayed in an SPSS Viewer window that is already open). You don't need this information for what you are doing here, so you can just close the window.

2. **Choose Data⇨Define Variable Properties.**

   The Define Variable Properties dialog box appears.

3. **On the left, select all the names of the variables you want to check, and click the triangle in the center of the window to move them to the right, as shown in Figure 4-12.**

4. **Click the Continue button.**

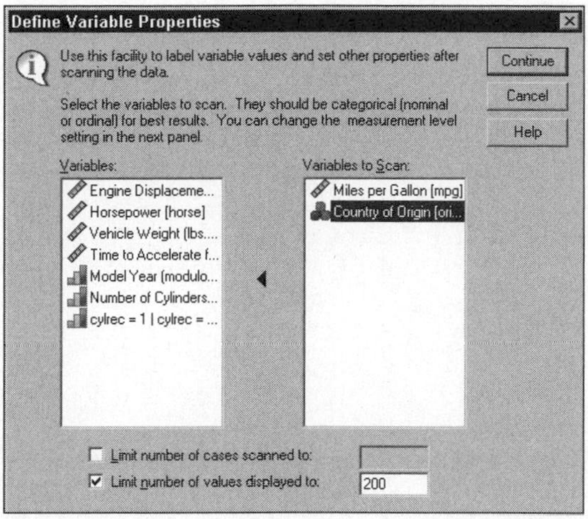

Figure 4-12:
Selecting
variables to
check their
properties.

5. **Select one of the variable names in the list on the left.**

   Its different values appear in the center of the window, as shown in
   Figure 4-13. In this example, all the values have a name assigned to them.

6. **In the top center of the window, ask SPSS to suggest a new type for
   this variable.**

   To do so, click the Suggest button. The window in Figure 4-14 appears,
   telling you what SPSS concludes about this variable and its values. This
   same window, with different text, will appear for each variable you test.
   Sometimes the text will suggest changes in the variable definition, and
   sometimes it will not.

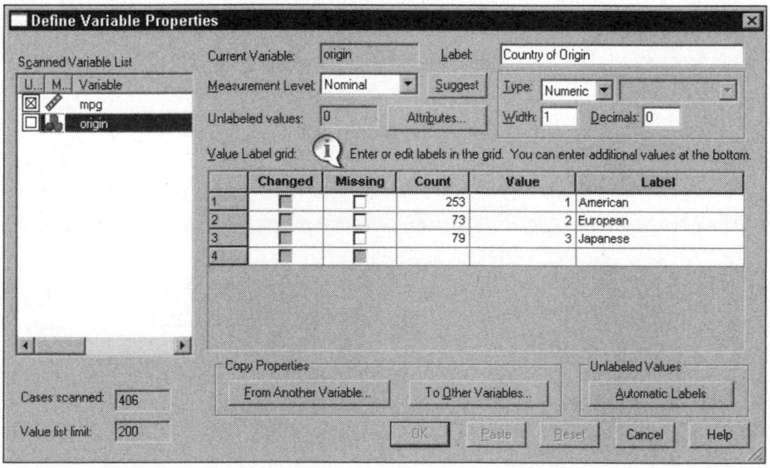

Figure 4-13:
The values
of the
selected
variable.

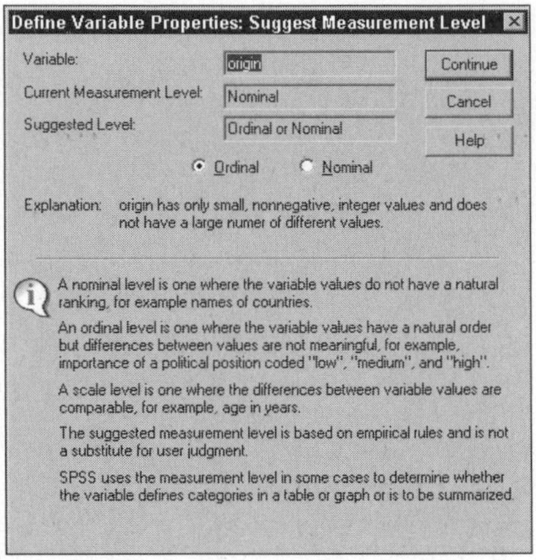

Figure 4-14:
SPSS
concludes
from the
pattern of
values
whether you
may have
chosen the
wrong mea-
surement.

7. **To apply any changes, click Continue.**

   You return to the window shown in Figure 4-13, where you can select another variable.

You won't want to make changes to all your variables, but SPSS will help you find the ones that you do need to change. Values defined as Missing are not included in the computations. The text in the window always explains the criteria used to reach a conclusion, and SPSS allows you to make the final decision.

Chapter 5

Reading and Writing Files

*T*here is no need to put your data into the computer more than once. If you've entered your data in another program, you can copy it from there into SPSS because every program worth using has some form of output that can serve as input to SPSS.

The SPSS File Format

SPSS has its own format for storing data, and several example files in this format are copied to your computer as part of the normal SPSS installation. These files have the `.sav` extension and are in the same directory as your SPSS installation. You can load any one of them by choosing File⇨Open⇨ Data and selecting the file to be loaded. The variable names and data will be loaded and will fill your SPSS window.

If you have SPSS filled with data, you can save it to a disk file by choosing File⇨Save As and providing a name for the file. Or if you've loaded the information from a file, or have previously saved a copy of the information to a file, you can simply use the File⇨Save selection to overwrite the previous file with a fresh copy of what you have — both variable definitions and data. This file contains special codes and can't be used to export your data to another application. This file format is only for saving SPSS data that you want to read back into SPSS at a later time.

You can be fooled by the way SPSS help uses the word *file*. If you have defined data and variables in your program, the SPSS documentation often refers to it all as a file, even though it may have never been written to disk. They also refer to the item written to disk as a file, so watch the context.

When you write your file to disk, if you don't add the .sav (or .SAV) extension to the file name, SPSS adds it for you. When you use File⇨Load⇨Data to display the list of files, you may or may not see the extension on the file name (it depends on how your Windows system is configured), but it's there.

Formatting a Text File for Input into SPSS

If your data is in an application that can't directly create a file of a type that SPSS can read, getting the data into SPSS may be easier than you think. If you can get the information out of your application and into a text file, it's fairly easy to have SPSS read the text file. However, some applications are more obliging than others when it comes to writing the information to disk. Look for an Export menu option — it usually has some options that will allow you to organize the output text in a form you want. (A description of possible organizations is coming up.)

If the application doesn't allow you to format text the way you want, maybe you can redirect printer output to a disk file and work from there. If you use the application's printer output, you may need to use your word processor to clean up the form of the data. I know this multistep operation sounds like a lot of work, but I'm not the one who put my data into that thing.

The text file does not need to include the variable names, just the values that go into the variables. Always save this kind of raw data as simple text; the file you store it in should have the .txt (or .TXT) extension so SPSS can recognize it for what it is.

You can format the data in the file by using spaces, tabs, commas, or semicolons to separate data items. Such dividers are known as *delimiters.* Actually, you don't have to separate the individual data items, but that requires that all data items be a specific length, because you have to tell SPSS exactly how long each one is.

The most intuitive format is to have one case (one row of data) per line of text. That means the data items in your text file are in the same positions they will be in when they are read into SPSS. Alternatively, you can have all your data formatted as one long stream, but you will have to tell SPSS how many items go into each case.

Reading Simple Data from a Text File

This section contains an example of a procedure you can follow to read data from a simple text file into SPSS. The file is a simple file named garbler.txt. It contains two cases (rows of data) as two lines of text, with the data items in the two lines separated by spaces. The content of the file is as follows:

```
"Pat" 1 35 3.00 9
"Chris" 1 22 2.4 7
```

The following example reads this text file and inserts it into the cells of SPSS. Along the way, SPSS will keep you informed about what's going on so there won't be any big surprises at the end.

1. **Choose File⇨Read Text Data.**

   The file selection window shown in Figure 5-1 appears.

2. **Select the `garbler.txt` file, and then click the Open button.**

   The screen shown in Figure 5-2 appears, for loading and formatting your data.

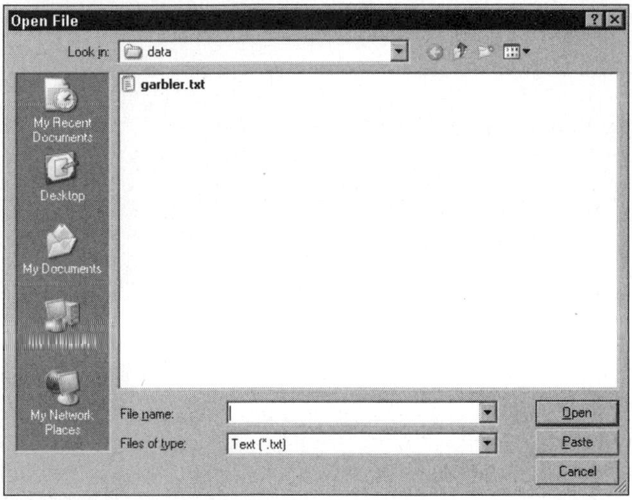

Figure 5-1: Locate the file you want to read.

3. **Examine the input data.**

   The screen lets you peek at the contents of the input file so you can verify that you've chosen the right file. Also, if you have a predefined format (which we don't, in this example), you can select it here and skip some of the later steps. If your data doesn't show up nicely separated into values the way you want, you may be able to correct it in a later step. Don't panic just yet.

4. **Click the Next button.**

   The screen shown in Figure 5-3 appears.

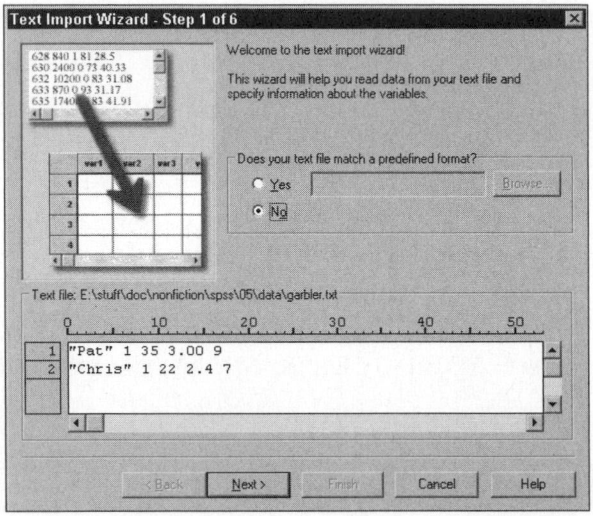

Figure 5-2:
Make
certain your
data looks
reasonable.

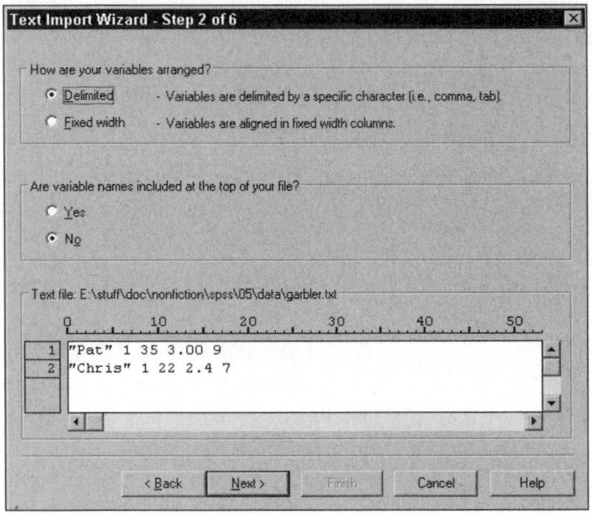

Figure 5-3:
Specify
whether the
fields are
delimited
and
whether the
variable
names are
included.

5. Specify how your data is delimited.

As you can see in this example, SPSS takes a guess, but you can also specify how your data is organized. It can be divided using spaces (as in this example), commas, tabs, semicolons, or some combination. Or your data may not be divided — it may be that all the data items are jammed together and each has a fixed width. If your text file includes the names of the variables (I'll show you how this works in a minute), you need to tell SPSS.

6. Click the Next button.

The screen shown in Figure 5-4 appears.

7. Specify how SPSS is to interpret the text.

You can tell SPSS something about the file and which data you want to read:

- Perhaps some lines at the top of the file should be ignored — this happens when you're reading data from text intended for printing and header information is at the top. By telling SPSS about it, those first lines can be skipped.

- Also, you can have one line of text represent one case (one row of data in SPSS), or you can have SPSS count the variables to determine where each row starts.

- And you don't have to read the entire file — you can select a maximum number of lines to read starting at the beginning of the file, or you can select a percentage of the total and have lines of text randomly selected throughout the file. Specifying a limited selection can be useful if you have a large file and would like to test parts of it.

8. Click the Next button.

The screen shown in Figure 5-5 appears.

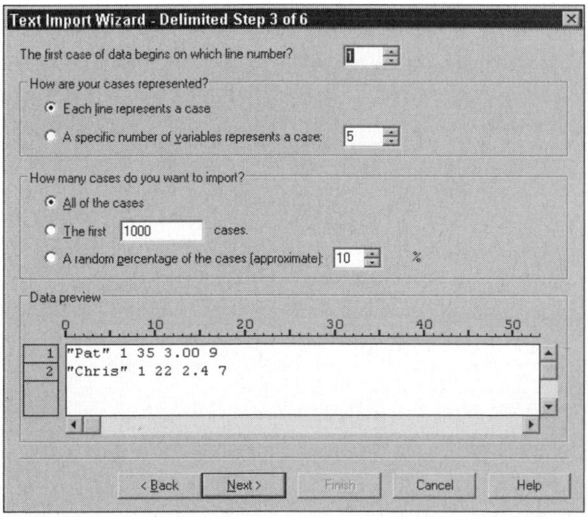

Figure 5-4:
Specify where the data appears in the file.

Text Import Wizard - Delimited Step 4 of 6

Which delimiters appear between variables?
☐ Tab   ☑ Space
☐ Comma   ☐ Semicolon
☐ Other: [   ]

What is the text qualifier?
○ None
○ Single quote
● Double quote
○ Other: [   ]

Data preview

V1	V2	V3	V4	V5
Pat	1	35	3.00	9
Chris	1	22	2.4	7

[< Back] [Next >] [Finish] [Cancel] [Help]

Figure 5-5: Specify the delimiters that go between data items and which quotes to use for strings.

9. **Specify the delimiters to use between data.**

SPSS knows how to use commas, spaces, tabs, and semicolons as delimiting characters. You can even use some other character as a delimiter by selecting Other and then typing the character into the blank. You can also specify whether your text is formatted with quotes (as in our example) and whether you use single or double quotes. Strings must be surrounded in quotes if they contain any of the characters being used as delimiters.

You can specify that a data item is missing in your text file. Simply use two delimiters in a row, without intervening data.

10. **Click the Next button.**

11. **Specify each variable name and type.**

SPSS assigns the variables the names V1, V2, V3, and so on. To change a name, select it in the column heading at the bottom of the window, and then type the new name in the Variable Name field at the top. You can select the format from the Data Format pull-down list, as shown in Figure 5-6. If you need to refine your data types and whatnot, you can do so later. The point here is to get the data into SPSS.

12. **Click the Next button.**

The screen shown in Figure 5-7 appears.

13. **Decide whether you want to save the information about the file format for future use.**

This is something you would do if you'll be loading more files of this same format into SPSS — it reduces the number of questions to answer and the amount of formatting to do next time. You also have the chance to grab a copy of the Syntax Language instructions that do all this, but unless you know about the Syntax Language (as described in Chapters 15 and 16), it's best to pretend that the option doesn't exist. The Cache option is a bit odd. I don't know why it's there, unless SPSS has some problem with huge files. SPSS seems to load data faster with it than without it, but it's strictly an internal thing and SPSS works just fine either way.

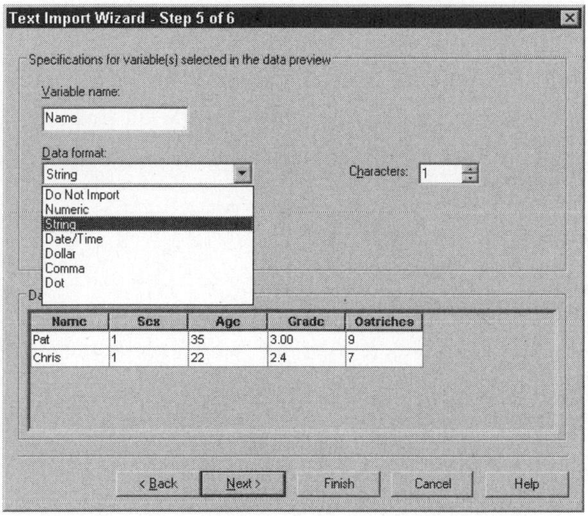

Figure 5-6:
Name your variables and select their data types.

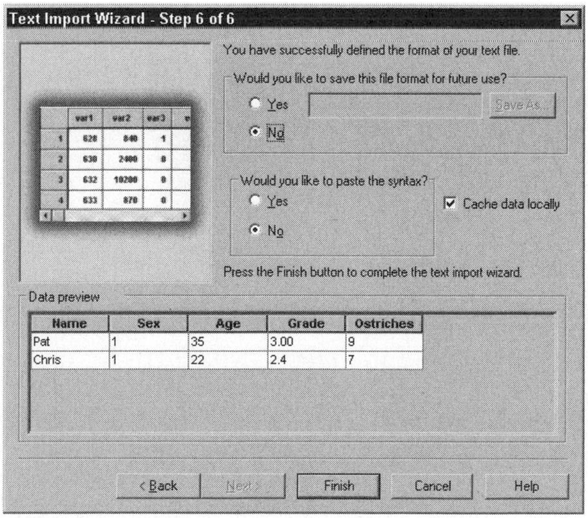

Figure 5-7:
Save the format, grab the syntax, or enable caching.

14. **Click the Finish button.**

    Depending on the type of data conversions and the amount of formatting, SPSS may take a bit of time to finish. But be patient, and the SPSS Data View window will eventually display your data.

15. **Look at the data. Correct your data types and formats, if necessary. Then save it all to a file by choosing File⇨Save As.**

    You are instructed to enter a file name. You can just call it `garbler`. The new file will have the `.sav` extension, which indicates that it's a standard SPSS file.

The SPSS way of reading data is a lot more flexible than this simple example demonstrates. Here, a file named `headgarbler.txt` is that same data formatted slightly differently:

```
Name Sex Age GradePoint Ostriches
Pat,1,35,3.00,9,Chris,1,22,2.4,7
```

This time, the data in the file is preceded by the variable names listed on the first line, the data is all in one long line, and the data is separated by commas. To read this into SPSS, you start the same way you did before. However, SPSS can't figure it all out in step 1 this time, as shown in Figure 5-8. SPSS can't even tell which is header and which is data.

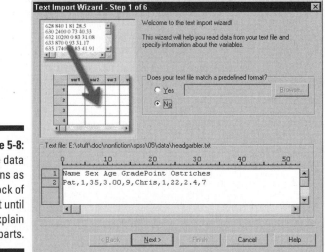

Figure 5-8: The data remains as a block of text until you explain the parts.

In step 2 of 6, you select the option that informs SPSS that the variable names appear in the first line of text. Then in step 3 of 6, as shown in Figure 5-9, you specify that the data begins on line 2 of the text file. It's possible for the data to begin several lines down in the input text file, but if variable names are present, they must be on the first line. Also, when you specify variable names, SPSS ignores the beginning and ending of lines, and counts the data values to determine when it has a complete row (case).

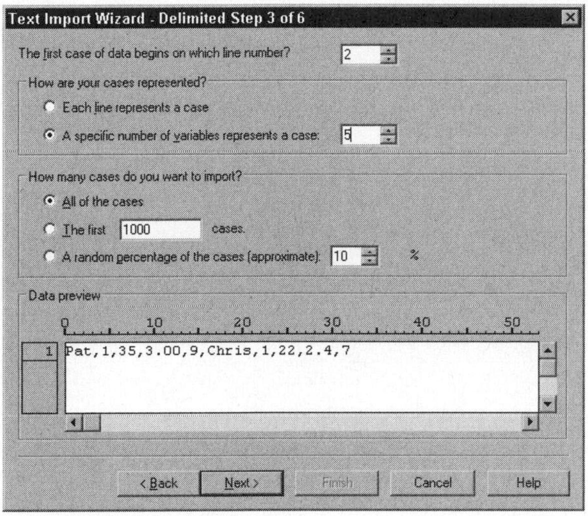

Figure 5-9:
Specify that
the data
starts on
line 2 and
each case
has 5 data
items.

In step 4 of 6, shown in Figure 5-10, commas and spaces were chosen as delimiters. (Although no spaces appear in the data in this example, it doesn't hurt to include a space delimiter if it may occur somewhere in your data.) Also, None was chosen for the characters surrounding string values. In this example, SPSS figured the spacing out on its own and used these settings for its default. Also, by the time you reach step 4 of 6, SPSS has started organizing the data according to your definitions. It has already read the variable names and included them as column headers.

In step 5 of 6, you have the opportunity to change the variable names and specify their types. Here again, you see that SPSS has made a guess for the type of each one.

After you complete step 6 of 6, click the Finish button and wait for the data to load, as shown in Figure 5-11.

You can see who has how many ostriches, but you still have a little work to do. For example, switch to Variable View, change the sex variable to a nominal data type, and assign the names "male" and "female" to the values 1 and 2. (You can't assume anything about sex by the names.) You might want to add some descriptive labels. For example, the variable named "ostriches" could be given the descriptive name "ostrich count in front yard". See how a good descriptive name can clear up a little mystery?

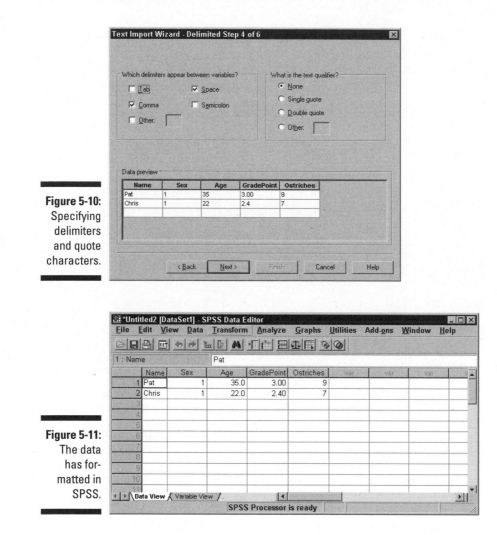

Figure 5-10:
Specifying delimiters and quote characters.

Figure 5-11:
The data has formatted in SPSS.

Transferring Data from Another Program

You can get your data into SPSS from a file created by another program, but it isn't always easy. SPSS knows how to read some file formats, but if you're not careful you'll find your data stored in an odd file format, and deciphering some file formats can be as confusing as working Klingon trigonometry. SPSS can read only from file formats it knows.

Reading from an unknown program type

You can often use copy-and-paste selections to transfer data from another application into SPSS, but that method has its drawbacks. The places you're copying from and to are usually larger than the screen, so highlighting and selecting can be tricky. You must be ready to choose Edit⇨Undo when necessary.

A better method is to write the data to a file in a format understood by SPSS, and then read the file into SPSS. SPSS knows how to read some file formats directly. Using such a file as an intermediary means you have an extra backup copy of your data, and that's never a bad idea.

Reading an Excel file

SPSS knows how to read Excel files directly. If you want to read the data from an Excel file, I suggest you read the steps in "Reading Simple Data from a Text File," earlier in this chapter, because the two processes are similar. If you understand the decisions you have to make in reading a text file, reading from an Excel file will be duck soup. Figure 5-12 shows the appearance of data displayed by Excel.

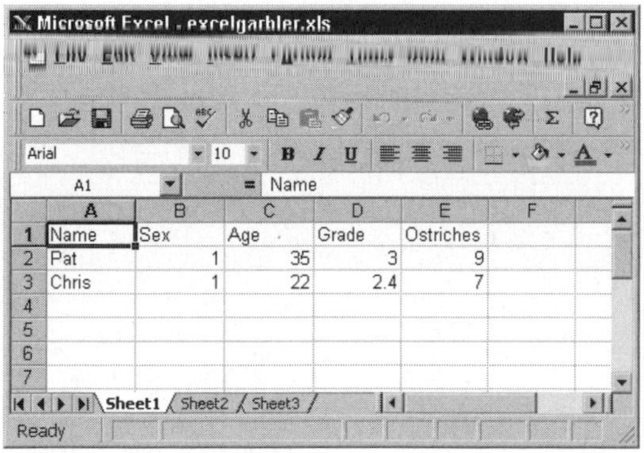

Figure 5-12: A simple example of Excel spreadsheet data.

Do the following to read this data into SPSS:

1. Save the Excel data to a file.

In this example, the file is called `excelgarbler.xls`. If you want to copy only a portion of the spreadsheet, make a note of the cells in the upper-left and lower-right corners of the group you want.

2. **Close Excel.**

   You must stop the Excel program from running before you can access the file from SPSS.

3. **Choose File➪Open➪Data.**

4. **Select the .xls file type, as shown in Figure 5-13, and then click Open.**

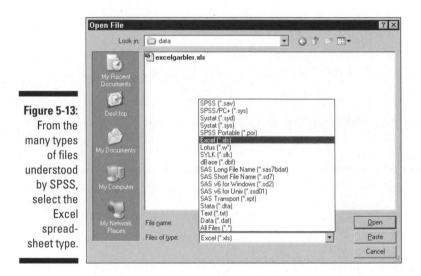

Figure 5-13: From the many types of files understood by SPSS, select the Excel spreadsheet type.

5. **Select the data to include.**

   An Excel file can contain more than one sheet, and you can choose the one you want from the pull-down list, as shown in Figure 5-14. Also, if you've elected to read only part of the data, enter the Excel cell numbers of the upper-left and lower-right corners here. You specify the range of cells the same way you would in Excel — using two cell numbers separated by a colon. Don't worry about the maximum length for strings.

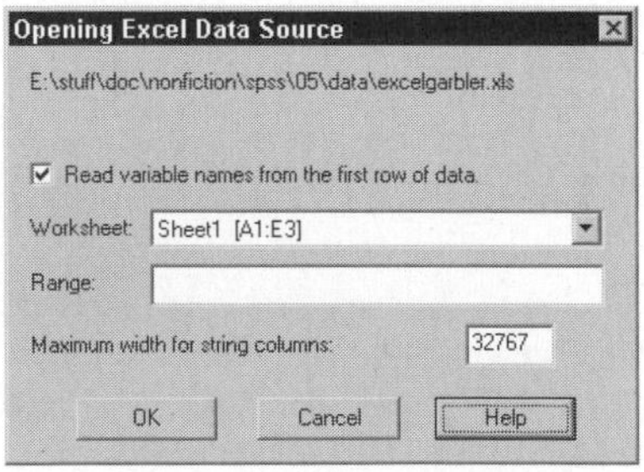

Figure 5-14: Select which data in the spreadsheet to include.

6. **Click OK.**

   Your data appears in the SPSS window.

7. **Check your variables and adjust their definitions as necessary.**

   SPSS makes a bunch of assumptions about your data, and it probably made some wrong ones. Closely examine and adjust your variable definitions by switching to Variable View and making the necessary changes.

8. **Save the file using your chosen SPSS name, and you're off and running.**

Reading from a known program type

SPSS recognizes the file formats of several applications. The preceding example — reading an Excel spreadsheet file — is just one of the types SPSS can work with. Following is a complete list:

- dBase (.dbf): An interactive database system
- Excel (.xls): Spreadsheet for performing calculations with numbers in a grid
- Lotus (.w): Spreadsheet for performing calculations with numbers in a grid
- SAS (.sas7bdat, .sdy, .sd2, .ssd, and .xpt): Statistical analysis software
- Stata (.dta): Statistical analysis and graphics software
- SYLK (.slk): A symbolic link file format for transporting data from one application to another
- SYSTAT (.syd and .sys): Software that produces statistical and graphical results

Although SPSS knows how to read any of these, you may still need to make a decision from time to time (such as with Excel data in the previous example, where you could select the sheet and the cell numbers). But you have some advantages. You know exactly what you want (the form of data appearing in SPSS is simple, and what you see is what you get), SPSS has some reasonable defaults and makes some good guesses along the way, and you can always fiddle with things after you've loaded them.

You are only reading from the other data, so you can't hurt it. Besides, you have everything safely backed up, don't you? If the process gets hopelessly balled up, you can always call it quits and start over. That's the way I do it — I think of it as my learning process.

Saving Data and Images

Writing data from SPSS is easier than reading data into SPSS. All you do is choose File⇨Save As, select your file type, and then enter a file name. You have lots of file types to choose from. You can write your data in two plain text formats, Excel spreadsheet format, three Lotus formats, three dBase formats, and six each of SAS and Stata formats.

If you'll be exporting data from SPSS into another application, find out what kinds of files the other application can read, and then use SPSS to write in one of those formats.

A second form of output from SPSS is an image. If you've generated a graphic that you want to insert into your word processor or place on your Web site, SPSS is ready to help you do it. I almost wish it were hard to do so I could look smart showing you how, but it isn't.

When you go through the steps to produce a graph, as explained in Part III, you'll be looking at the resulting graphics in SPSS Viewer, which is shown in Figure 5-15.

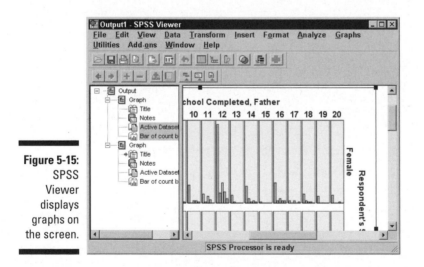

Figure 5-15:
SPSS
Viewer
displays
graphs on
the screen.

From SPSS Viewer, you can export images (and do some other things too):

1. **Produce a graph or table.**

   You can use any of the examples in Part III to produce a graphic display. SPSS Viewer pops up and displays the output.

2. **Choose File⇨Export.**

3. **In the Export pull-down list, select what you want to output, as shown in Figure 5-16.**

   Your choices are as follows:

   - Output Document: Outputs text and graphics into a single file

   - Output Document (No Charts): Outputs only text — the numbers from the table along with any text

   - Charts Only: Outputs only the annotated pictures — the graphics

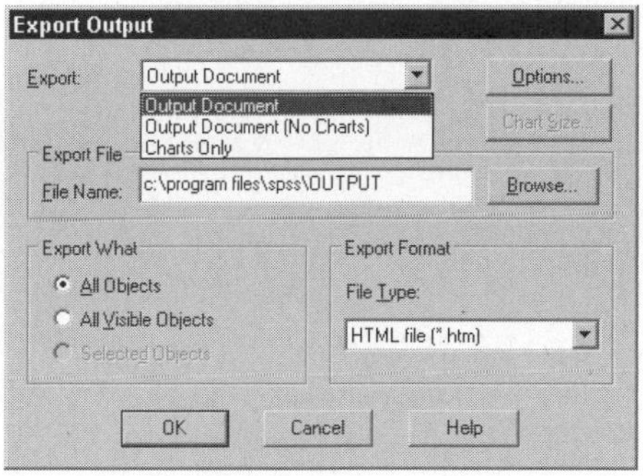

Figure 5-16: These selections control what gets output and into what format.

4. **In the Export What section, select which items to include in the output.**

   You can elect to have all objects output, all visible objects output, or only the ones you've selected. In Figure 5-15, for example, the panel on the left indicates that the items were selected for display. The *visibility* of an object refers to its name appearing in the list — if you collapse the list so that a name can't be seen, the item is not visible. You can select items by clicking them or by selecting their names in the list on the left.

5. **In the Export Format area, choose an output format.**

   Your choices will vary according to what you decided to output at the top of the window:

   - HTML files can be used for text both with and without graphics. If graphics are included, you need to export those separately and they will need to be compatible with the HTML links.

   - Text files can be output to include graphics. This is accomplished by a line in the text file naming the file for each graphic object. The listing isn't the graphic — it is only the name of a file containing the graphic. The graphic file is also written by SPSS.

   - Excel spreadsheet files can be written to include only the text or the text including the graphics, but not graphics alone.

   - Word documents are written in RTF (rich text format), which can be copied into a Word document. This works for text with or without graphics, but not for graphics alone.

- PowerPoint documents can be written as text or as text with graphics, but not as just graphics. The graphic portion of the text is written to disk in TIFF format.

- Image files without text can be written in a number of formats, including JPEG, TIFF, BMP, WMF, and the new PNG formats. They can also be written in the Macintosh PICT format, and the not-quite-so-common formats of Enhanced Metafile (EMF) and Postscript (EPS).

6. Select the directory and file name, and click Save.

Click the Browse button and you'll be able to select the directory and the name of the file you want to create. The Save button does not write the file — it only inserts your selected name into the Export Output window.

7. Click the OK button.

The file is written to disk in the chosen format at the chosen location.

Chapter 6

Data and Data Types

● ●

In This Chapter

▶ Understanding the special properties of dates and times

▶ Working with data that comes at regular intervals

▶ Creating multiple response sets

▶ Copying variable definitions from another file

● ●

*I*f you've worked on a table and have some variable definitions that are nifty, you can copy them into a new table (or even into an old table). Dates, times, and schedules are important in statistics but are usually hard to work with arithmetically. However, all you have to do is tell SPSS how you would like to handle them and all the hard calculating can be taken care of for you. Arithmetic that normally would be tedious and boring can be automated by assigning the appropriate data types.

Dates and Times

Calendar and clock arithmetic can be tricky, but SPSS can handle it all for you. You enter the date and time in whatever format you specify, and SPSS converts it into its internal form for calculations. Also, SPSS displays the date and time in your specified format, so it's easy to read.

SPSS understands the meaning of slashes, commas, colons, blanks, and names in the dates and times you enter, so you can write the date and time almost any way you'd like. If SPSS can't figure out what you've typed, it clears what you typed and waits for you to type it again.

Internally, SPSS keeps all dates as a positive or negative count of the number of seconds from a zero date. As a result, all dates also include the time of day. You can choose a format that includes or excludes the display of the time, but the information is always there. You can change the display format without loss of data. If the time is not included in your format, SPSS assumes zero hours and minutes (midnight) when you enter data.

You determine the data type for each variable in the Data View window. The type is chosen from the list of types shown in Figure 6-1. On the right, you select a format. SPSS uses this format to interpret your input and to format the dates for display.

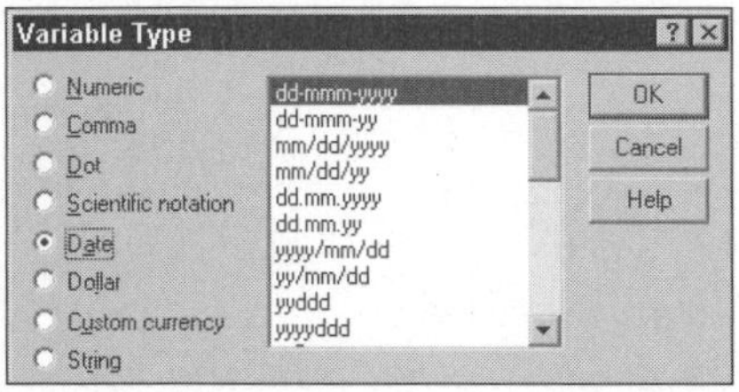

Figure 6-1:
Select the
data type
and the
format.

Variable Type	? X
○ Numeric	dd-mmm-yyyy / OK
○ Comma	dd-mmm-yy
○ Dot	mm/dd/yyyy / Cancel
○ Scientific notation	mm/dd/yy
● Date	dd.mm.yyyy / Help
○ Dollar	dd.mm.yy
○ Custom currency	yyyy/mm/dd
○ String	yy/mm/dd
	yyddd
	yyyyddd

SPSS uses the format you select for both reading your input and formatting the output of dates and times.

The Columns of the date variable in the Variable View is important. The column width determines the maximum number of characters that can be displayed, and if you choose a format that is too wide to fit, the date will show up only as a row of asterisks.

The available formats are defined as a group and change according to the variable type. For example, Dollar type will have a different list of choices than the Date type shown in Figure 6-1.

The list of format definitions you have to choose from are constructed by combining the specifiers listed in Table 6-1. Format definitions look like mm/dd/yy and ddd:hh:mm.

Data types

A data type is nothing more than the definition of what a number means. Without a definition, a number serves no purpose. For example, the number 3 could have entirely different meanings. It could be a number of miles, or one answer to a multiple-choice question, or the number of jelly beans in your left pocket. The data type is more than just a tag — it determines how the value can be manipulated. For example, 3 miles can also be written as 15,840 feet or as 24 furlongs. Some data types require special arithmetic. For example, if the number 50 represents the number of minutes past 2 o'clock, adding 15 to it will result in the number 5, but the new value represents the number of minutes past 3 o'clock.

Table 6-1	Specifiers in Date and Time Formats
Specifier	*Means*
dd	A two-digit day of the month in the range 01, 02, . . . , 30, 31.
ddd	A three-digit day of the year in the range 001, 002, . . . , 364, 365.
hh	A two-digit hour of the day in the range 00, 01, . . . , 22, 23.
Jan, Feb, . . .	The abbreviated name of the month of the year, as in JAN, FEB, . . . , NOV, DEC.
January, February, . . .	The name of the month of the year, as in JANUARY, FEBRUARY, . . . , NOVEMBER, DECEMBER.
mm	When adjacent to a dd specifier in a format, a two-digit month of the year in the range 01, 02, . . . , 11, 12. When adjacent to an hh specifier in a format, a two-digit specifier of the minute in the range 00, 01, . . . , 58, 59.
mmm	A three-character name of a month, as in JAN, FEB, . . . , NOV, DEC.
Mon, Tue, . . .	The abbreviated name of the day of the week, as in MON, TUE, . . . , SAT, SUN.
Monday, Tuesday, . . .	The name of the day of the week, as in MONDAY, TUESDAY, . . . , SATURDAY, SUNDAY.
q Q	The quarter of the year, as in 1 Q, 2 Q, 3 Q, or 4 Q.
Ss	Following a colon, the number of seconds in the range 00, 01, . . . , 58, 59. Following a period, the number of hundredths of a second.
ww WK	The one- or two-digit number of the week of the year in the range 1 WK, 2 WK, . . . , 51 WK, 52 WK. Note: Although week numbers can be either one or two digits, the numbers always line up when printed in columns because SPSS inserts a blank in front of single-digit numbers.
yy	A two-digit year in the range 00, 01, . . . , 98, 99. The assumed first two digits of the four-digit year this represents is determined by the configuration found at Edit⇨Options⇨Data.
yyyy	A four-digit year in the range 0001, 0002, . . . , 9998, 9999.

You can go back and change the format of a date variable at any time without fear of losing information. For example, you could enter the data under a format that accepted only the year, month, and day, and then change the format to something that contains only the hours and minutes. The format may not display all the information you entered (in fact, in this case, it won't), but when you change the format back to something more inclusive, you will find that all your data is still there.

To enter data, you should choose a format, any format, that contains all the data you have. You can later change to a more limited format that displays only the information you want. But you can't go the other way. If you later choose a format that doesn't leave parts out, you will see the defaults that were inserted by SPSS when you entered the data.

Time Schedule

Sometimes you have data that's gathered at regular intervals, and you need to know the time each data record was gathered. But interval tracking can be more than simple counting. For example, you might need to track information for each new case hourly, for an eight-hour workday, for five workdays each week, for a few months. This repetition pattern is known as the *periodicity* of the data. Now that's a word you should never try to say out loud in public until you've practiced in private.

Here's the good news. SPSS can not only create your periodicity variables but can also insert the periodic values into the variables for all your cases. To do all this, use the following steps:

1. **Define your variables and enter your data.**

   Do not define any of the periodicity variables — they will be generated later automatically. The other variables and data can be entered using any of the methods described in Chapters 4 and 5.

2. **In Variable View or Data View, choose Data⇨Define Dates.**

   The window shown in Figure 6-2 appears.

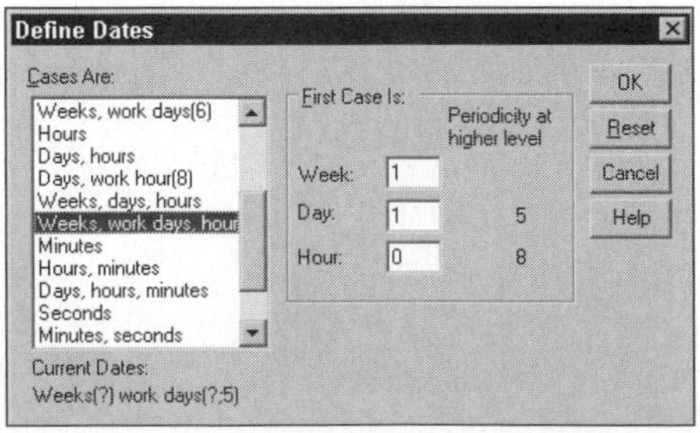

Figure 6-2: Select the desired periodicity for your data.

3. Select the desired periodicity.

The interval being defined in this example is once each hour, for an 8-hour day, for a 5-day work week. The starting week number is 1, the day number is 1, and the hour number is 0. The hour numbers count up to 8, and for each count of 8 hours, the day number increases by 1 until it reaches 5, then the week number increases by 1. Each time a number reaches its maximum, it starts over at the beginning.

4. Click the OK button.

You're done. The window shown in Figure 6-3 appears, listing the variables that have been defined and added to your previous definitions. The variable names end with an underscore character (_) to indicate that they have been generated automatically. You can close this window if you want — it's only informational.

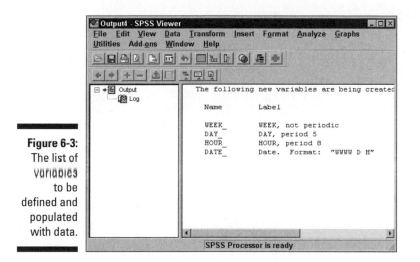

Figure 6-3:
The list of
variables
to be
defined and
populated
with data.

Figure 6-4 is the Variable View of the new variables that have been created. The variable named `score` already existed in the example. The new variables named `WEEK_`, `DAY_`, and `HOUR_` are numeric variables and are used to hold the numbers of the period. The `DATE_` variable is a string data type and holds a string representation of the value of the other three.

Switch to Data View and you see the screen shown in Figure 6-5. The first case was assigned the starting value for each of the new values, and each case was assigned the values for the next period in the sequence.

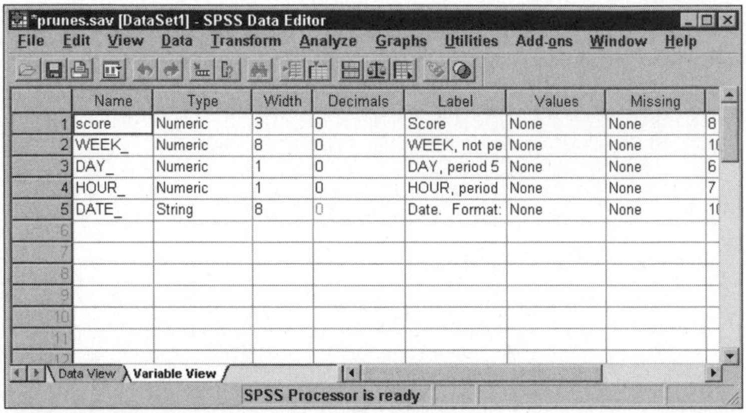

Figure 6-4:
The newly created variables are added to your table.

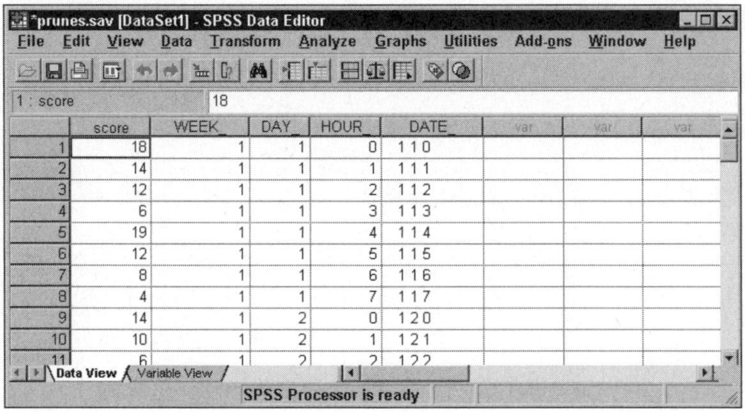

Figure 6-5:
The newly inserted values for the new variables.

Creating a Multiple Response Set

A *multiple response set* is very much like a new variable that is made out of other variables that you already have. A multiple response set acts like a variable in some ways, but in other ways it doesn't. You define it based on the variables you've already defined, but it doesn't show up in Variable View. It doesn't even show up when you list your data in Data View, but it does show up among the items you can choose from when defining graphs and tables.

The following steps explain how you can define a multiple response set, but not how you can use one — that will come later.

You can build a multiple response set based on two or more *dichotomy variables* or two or more *category variables*. For example, suppose you have two dichotomy variables with answers defined as 1 for no and 2 for yes. You can combine both into a multiple response set consisting of all the cases where the answer to both is yes, or the answer to both is no, or whatever combination you want.

Do the following to create a simple multiple response set:

1. **Create two dichotomy variables that both have 1 for no and 2 for yes as their possible answers, as shown in Figure 6-6.**

   You can do this with more than two variables, but they must all be of compatible types and contain the same set of possible values. The process of creating variables is described in Chapter 4.

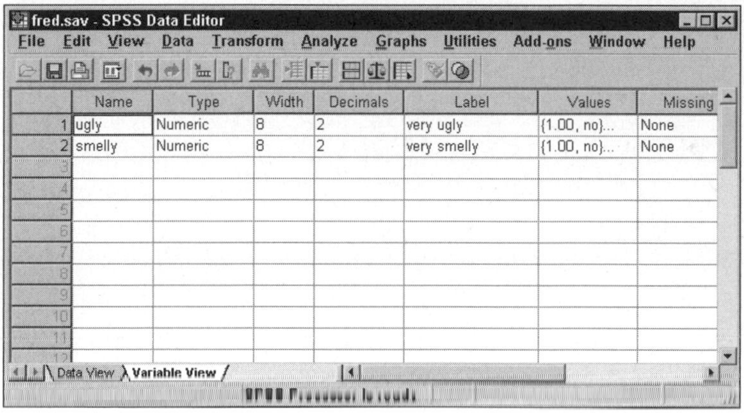

Figure 6-6:
The only two variables are nominals with possible values of yes and no.

2. **Choose Data⇨Define Multiple Response Sets.**

   The window shown in Figure 6-7 appears. Your variables appear in the Set Definition area. If you've already defined any multiple data sets, they appear in the list on the right.

3. **In the Set Definition list, select each variable you want to include in your new multiple data set, and then click the triangle to move the selections to the Variables in Set list.**

   You can move variable names back and forth until you get the list you want. In this example, we need both of them.

4. **In the Variable Coding area, select the Dichotomies option. Specify a Counted Value of 2.**

   With a Counted Value as 2, your new multiple response set will be a count of all the cases in which both variables have the value 2. That is, when you use the variable (for analysis or to draw a graph or whatever), it will only exist where both the dichotomy variables have the value 2. If you get a count of the number of occurrences of the variable, you will have a count of the cases in which the two base variables have a value of 2.

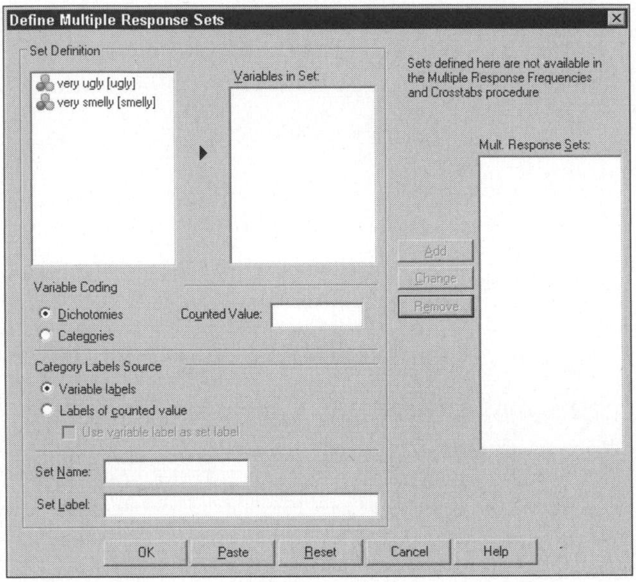

Figure 6-7: The window showing all the information about multiple response sets.

5. **Select a Set Name and (optionally) a Set Label.**

6. **Click OK.**

   The new multiple response set is created and a dollar sign ($) is placed before the name, as shown in Figure 6-8. The dollar sign identifies the name as that of a multiple response set variable.

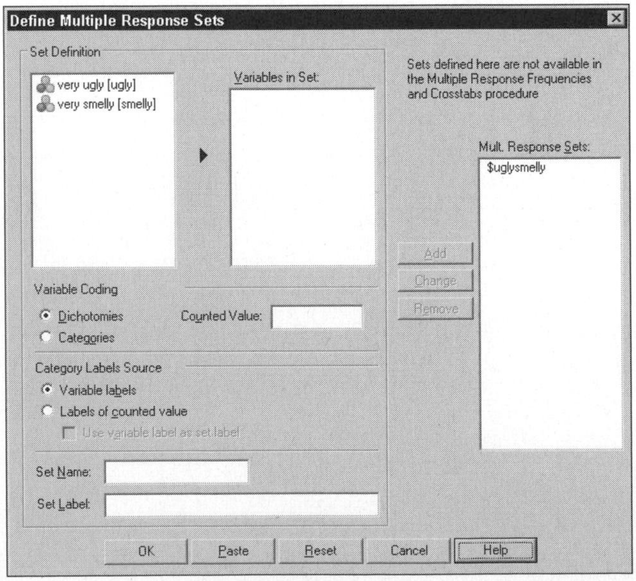

Figure 6-8: One response set has been defined.

This example used a pair of dichotomy yes/no variables, and built a set that counted the cases where they were both yes. But we could just as well use a group of category variables that all have the same set of answers. For example, if you have the variables Favorite Color, Car Color, and Underwear Color, you could create a multiple response set made up of a count of the instances where all three answers are Red.

Copying Data Properties

Suppose you have some data definitions in another SPSS file, and you want to copy one or more of those definitions but you don't want the data. SPSS enables you to choose from several files and to copy only the variable definitions you want into your current table.

If you have a variable *of the same name* defined in your table before you execute the copy, you will be able to change the existing variable definition by loading new information from another file. Otherwise, the copy will create a new variable.

The following steps show you how to copy data properties:

1. **Choose Data⇨Copy Data Properties.**

   The window in Figure 6-9 appears.

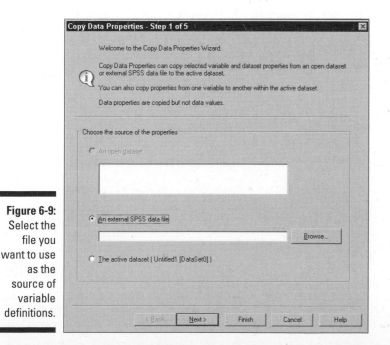

Figure 6-9:
Select the file you want to use as the source of variable definitions.

2. **Make certain the An External SPSS Data File option is selected.**

3. **Click the Browse button, locate the file from which you want to copy variable definitions, and then click Open.**

   The name of the selected file appears next to the Browse button.

4. **Click the Next button.**

5. **Select the variables you want.**

   Figure 6-10 displays the variable names that match in the source and destination. In the example, all three are selected, but you can turn the selection of each one on and off by holding down the Ctrl key and clicking the mouse on the one you want to select or deselect.

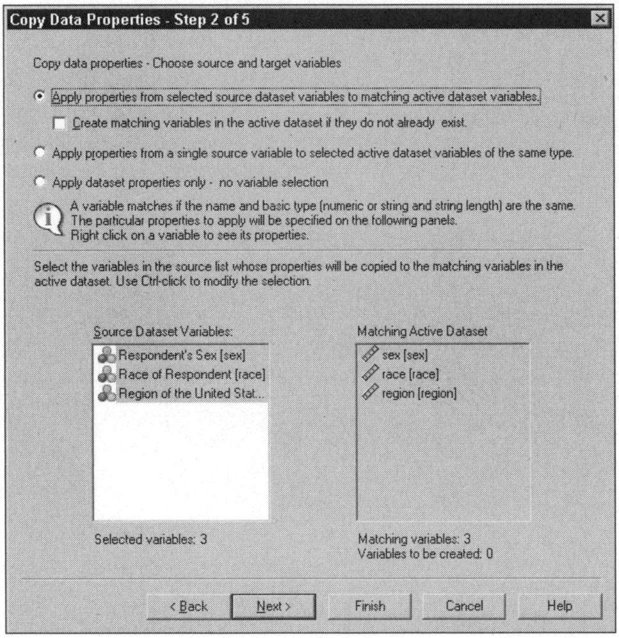

Figure 6-10:
Select the source variable names you want to use for definitions.

6. **To use the variables you have selected, click Next.**

   If you want to copy the complete definitions of all the variables you've selected and completely overwrite what you have, you can click the Finish button in this window. The Next button, as in this example, allows you to be more specific about which parts of the variable definitions you want to copy.

7. **Choose just what it is about the existing variable definitions you want to copy to the variables you're modifying.**

   In Figure 6-11, everything is selected by default, but you can skip any parts you don't want by deselecting them. These selections apply to all the variables you have chosen. If you want to handle each variable separately, you will need to run through this entire procedure again, selecting different variables each time.

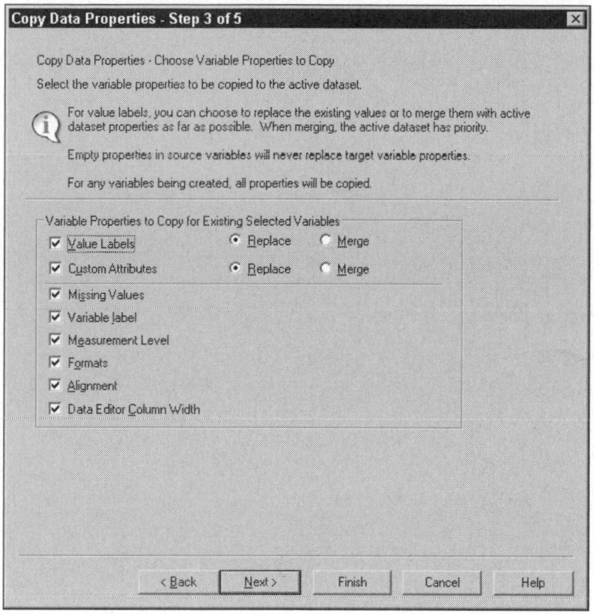

Figure 6-11:
Select
which
attributes
you want to
copy.

8. **Click Next to be able to select from a list of variable properties.**

   If you're satisfied with your choices, you can click the Finish button in this window to complete the process. Clicking Next, as in this example, makes it possible for you to select from a list of available properties to be copied.

9. **Choose any properties made available to you in the dialog box shown in Figure 6-12.**

   Depending on the variable type, different properties are available to be copied. As shown in Figure 6-12, the properties not available appear grayed out. By default, none of them are selected.

10. **Click Next to move to the final dialog box.**

    As shown in Figure 6-13, the screen displays the number of existing variable definitions to be changed, the number of new variables to be created, and the number of other properties that will be copied. You can elect to have the action take place immediately or have the set of instructions saved as a Command Syntax script so you can execute them later. Use of the Command Syntax language is described in Part V.

11. **Decide whether to execute the commands now or later.**

    You can click Finish to have the copy execute immediately.

12. Click Finish.

With the basic variable types and the property descriptions you can add to them, you should be able to concoct any type of variable you need.

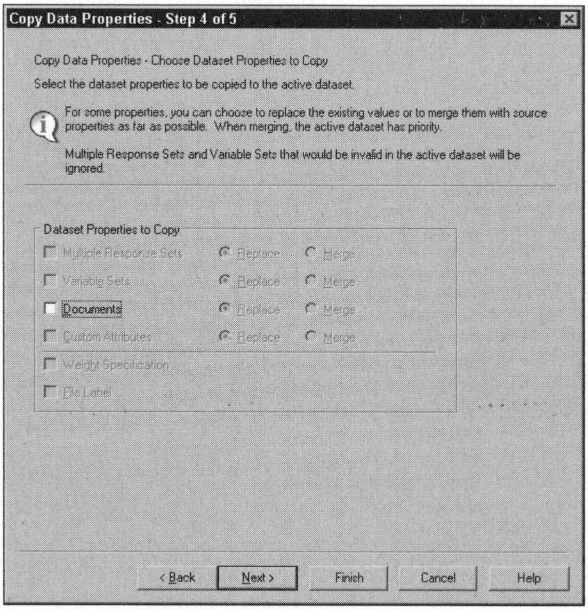

Figure 6-12:
Attributes
other than
variable
definitions
can be
copied from
the source.

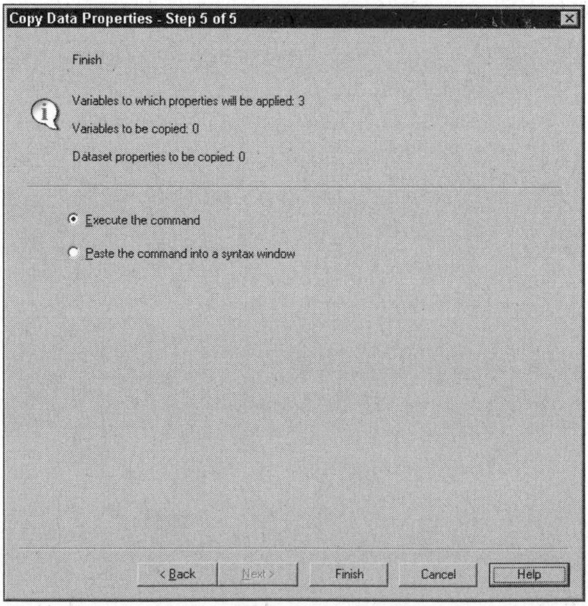

Figure 6-13:
Choose to
execute the
commands
or save the
commands
for later
execution.

Chapter 7

Messing with the Data
After It's in There

Aﬅer you get your raw data into SPSS, you may find that certain types of analyses are clumsy. You can make modifications to your data to put it in forms easier to work with — or, maybe not easier to work with, but easier to read and see what you have. This chapter contains some methods you can use to change your data without loss of information.

Sorting Cases

You can sort your cases (rows) so they appear in just about any order you want. The sorting is based on the values you entered for variables. The following example uses one of the data files that installs with SPSS. The data will be sorted with all males listed first, with the youngest males first within that sort order. These two variables — sex and age — are known as the *primary* and *secondary* sort keys.

You don't need to limit your sorting to two sort keys. You can have a third and fourth key, if necessary, but the later keys come into effect only when the keys that come before them hold identical values. In most cases, two sort keys are plenty to get what you want.

You can sort based on any variables, of any type, by simply selecting the variables as keys. For example:

1. **From the main menu, choose File⇨Open⇨Data and load the 1991 U.S. General Social Survey file, which is in the SPSS directory.**

   The result is the presentation of a collection of apparently unsorted cases shown in Figure 7-1.

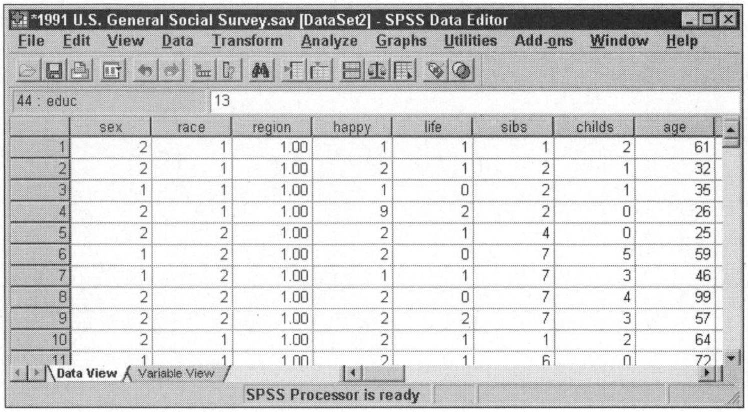

Figure 7-1: The data unsorted, as it is loaded directly from the data file.

2. **From the Data Editor window, choose File⇨New⇨Syntax, and the Syntax Editor window appears.**

3. **Enter the four words** SORT CASE SEX AGE. **as shown in Figure 7-2.**

   This is one line of Command Syntax language. Be sure to include the period at the end. Although the command will work without it, SPSS will complain.

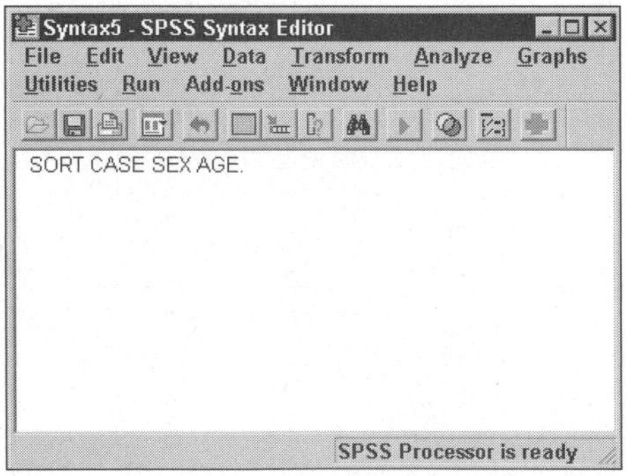

Figure 7-2: The Syntax Editor window containing a simple syntax.

4. From the main menu of the Syntax Editor window, choose Run⇨To End.

Close the Syntax Editor window, if you want, to look at the Data View tab of the SPSS Editor window, as shown in Figure 7-3. The data has been sorted with the male sex — represented by the number 1 — and the youngest age — which is 18 — at the top of the list of cases. It came up male first because male is 1, which is a smaller number than the 2 that represents female.

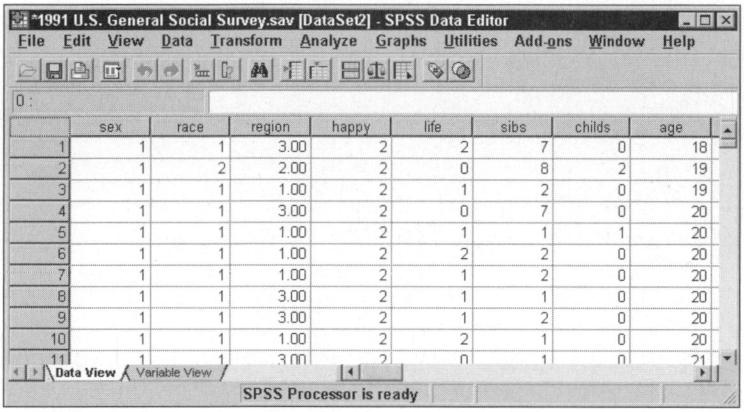

Figure 7-3:
The data sorted with the case of the youngest male first.

5. To change the order in which things are sorted, replace the command in the Syntax Editor window with SORT CASE SEX (D) AGE.

You can reverse the sort order for any or all sort key variables. The default is ascending order — smallest to largest — but you can specify descending order by following a variable name with a (D) indicator. The resulting sort, with the youngest female first, is shown in Figure 7-4.

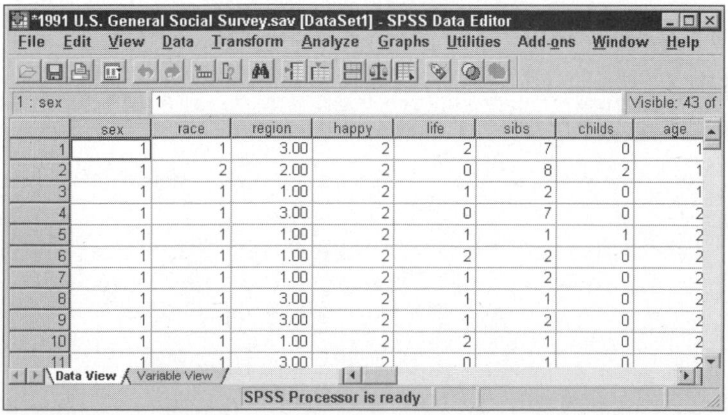

Figure 7-4:
The data sorted with the youngest female first.

Sorting data is strictly for the way you want it to appear in the table. The order of the data never affects the analysis.

The order of the sort keys is important. In the preceding example, if AGE had been chosen as the first key and SEX as the second, all 18-year-olds would have come up first in the list, and they would have been ordered by female and then male. Following that, the next age would have come up, and it would also have been ordered by sex. And so on.

If you want to change only the direction of a sort — ascending instead of descending or vice versa — you must also make a change in the sort key. That is, you need to perform a sort with a different key selection. If you don't, SPSS will not notice your change and will not perform the sort.

Using an ID to Identify Cases

Because cases can be sorted into different orders, and because the identifying numbers built into SPSS do not change position along with the cases, you may need to add an identifier to each case. To do this, you add a variable that contains the identifier, which can be a name, a date, or anything else.

When creating an identifier, you should probably name it id because some of the more advanced capabilities of SPSS look for it by that name. If you name it something else, SPSS will not automatically find it.

Probably the most common form of identifier is a simple number. The following example shows how an identification number is used to track employees and keep their records straight even though the order is changed:

1. **Choose File⇨Open⇨Data and open the Employee data.sav file.**

   The file is in the same directory you used to install SPSS. When the file loads, SPSS looks like the window in Figure 7-5. The values in the variable named id are the identifying numbers for the cases, and each case is the data for one employee. Numbers are used instead of names for identifiers, but employee names could be used as well.

2. **Sort the data, using the salary as the sort key and selecting descending (large to small) order.**

   To sort the data, use the technique described earlier in the chapter. When you do, the screen looks like Figure 7-6. The rows have all been re-ordered and the first column contains the identifying number of each case.

In this example, we used numbers as an identifier. Unless an order to the identifiers is important, you could use names instead.

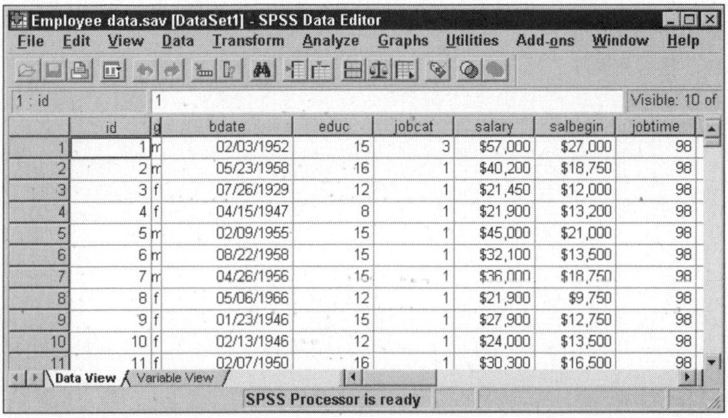

Figure 7-5:
Employee
information
with an
identifying
number
for each
record.

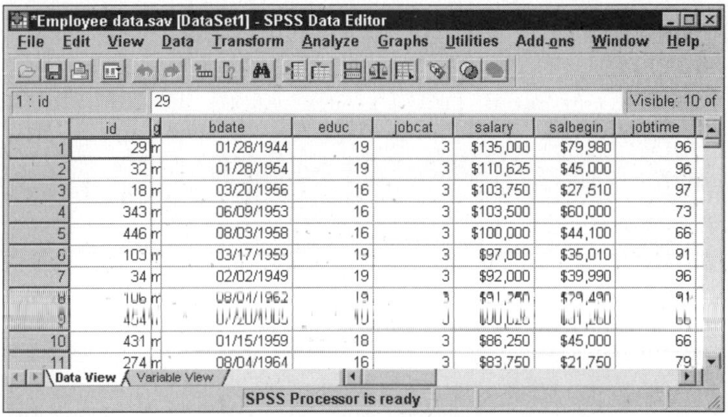

	id	g	bdate	educ	jobcat	salary	salbegin	jobtime
1	1	m	02/03/1952	15	3	$57,000	$27,000	98
2	2	m	05/23/1958	16	1	$40,200	$18,750	98
3	3	f	07/26/1929	12	1	$21,450	$12,000	98
4	4	f	04/15/1947	8	1	$21,900	$13,200	98
5	5	m	02/09/1955	15	1	$45,000	$21,000	98
6	6	m	08/22/1958	15	1	$32,100	$13,500	98
7	7	m	04/26/1956	15	1	$36,000	$18,750	98
8	8	f	05/06/1966	12	1	$21,900	$9,750	98
9	9	f	01/23/1946	15	1	$27,900	$12,750	98
10	10	f	02/13/1946	12	1	$24,000	$13,500	98
11	11	f	02/07/1950	16	1	$30,300	$16,500	98

Figure 7-6:
Employee
information
in descend-
ing salary
order
with an
identifying
number.

	id	g	bdate	educ	jobcat	salary	salbegin	jobtime
1	29	m	01/28/1944	19	3	$135,000	$79,980	96
2	32	m	01/28/1954	19	3	$110,625	$45,000	96
3	18	m	03/20/1956	16	3	$103,750	$27,510	97
4	343	m	06/09/1953	16	3	$103,500	$60,000	73
5	446	m	08/03/1958	16	3	$100,000	$44,100	66
6	103	m	03/17/1959	19	3	$97,000	$35,010	91
7	34	m	02/02/1949	19	3	$92,000	$39,990	96
8	106	m	08/04/1962	19	3	$91,250	$39,490	91
9	454		07/20/1965	10	3	$88,625	$31,260	66
10	431	m	01/15/1959	18	3	$86,250	$45,000	66
11	274	m	08/04/1964	16	3	$83,750	$21,750	79

Counting Case Occurrences

If your data is being used to keep track of multiple similar occurrences, you can automatically generate a count of the occurrences for each case. SPSS automates the process of creating a new variable and counting the values for you. You specify what value or values cause a variable to qualify, and SPSS counts the number of qualifying variables from among those you choose. You must have a number of variables that all normally take the same range of values. For example, if you have a number of expenses for each case, you could have SPSS count the number of expenses that exceed a certain threshold.

In the following example, people are listed as subscribers or nonsubscribers to three magazines, which are named simply mag1, mag2, and mag3. The following steps generate a total of the number of subscriptions for each person:

1. **Choose Open➪File➪Data and open the `magazines.sav` file.**

   The screen shown in Figure 7-7 appears.

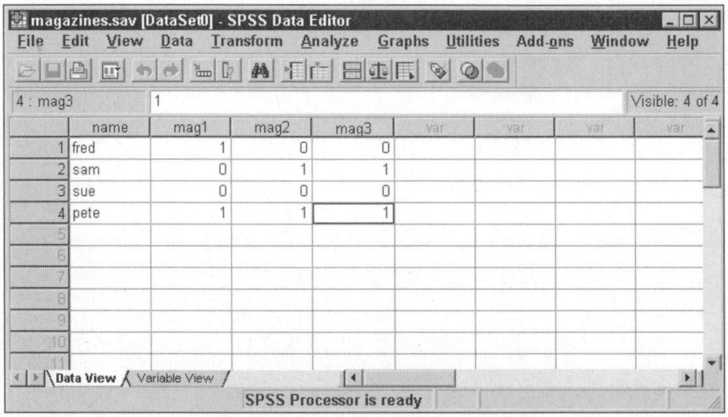

Figure 7-7: Each magazine has the value 1 for a subscriber and 0 for a nonsubscriber.

2. **Choose Transform➪Count Values Within Cases.**

   The screen shown in Figure 7-8 appears.

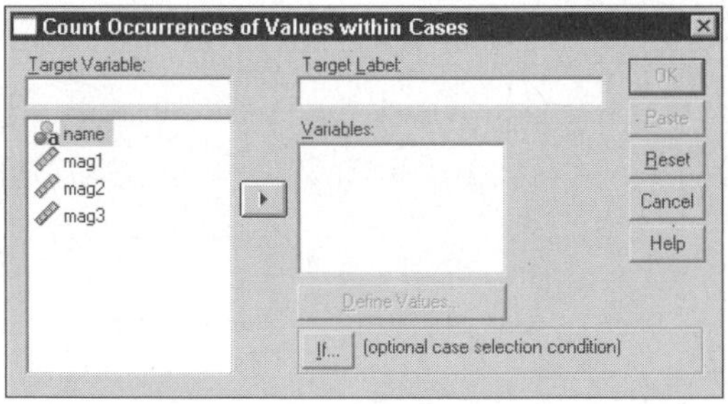

Figure 7-8: The initial value counting window.

3. **Select the name of each variable you want to use in the count, and then click the button marked with a triangle to move them from the panel on the left to the panel on the right labeled Variables.**

   This operation works only with numerics because it must perform calculations with the values. If you want, you can come up with both a name and a label to be assigned to the variable that will be created. In this example, the name is `count` and the label is `Count of subscriptions`, as shown in Figure 7-9.

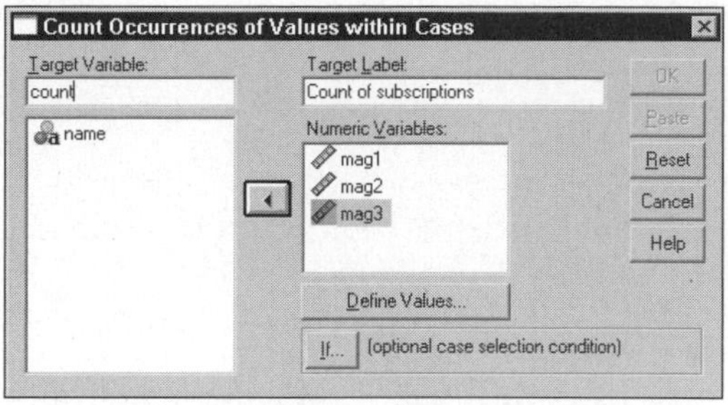

Figure 7-9:
The chosen variables to count, and the name of the new variable.

4. Click the Define Values button.

The window shown in Figure 7-10 appears. In this window, I've decided to count, from among the selected variables, those with the numeric value of 1, which in our example is the value signifying a subscription.

As you can see in the figure, the total can also be based on missing values and ranges of values. In the ranges, you can specify both the high and low values, or you can specify one end of the range and have the other end be either the largest or the smallest value in the set. The fact is that you can select a number of criteria and SPSS will check each variable against all of them.

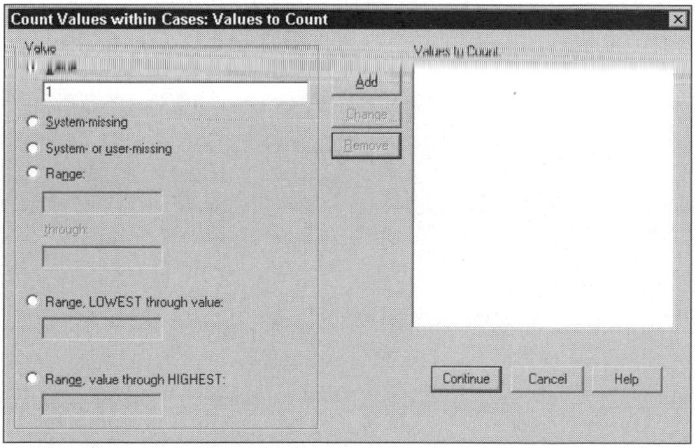

Figure 7-10:
Define the criteria by which variable inclusion in the count is determined.

5. Select a criterion you want to use and then click the Add button to move it to the panel on the right labeled Values to Count. Repeat as needed to define all your criteria.

6. Click the Continue button.

You return to the Count Occurrences of Values within Cases screen (refer to Figure 7-8).

7. **Click the If button.**

   The window shown in Figure 7-11 appears.

8. **Define your expression.**

   By default, all cases are included, but you can specify criteria here to exclude some cases. To do so, click the Include If Case Satisfies Condition option, and only the cases where the expression is true are considered as candidates for a count greater than 0. You can use any of the variables in the expression. And by using the number pad, the operator buttons, and the function selection, you can construct any expression you want. For more information on constructing expressions, see Part V.

9. **Click the Continue button to have SPSS accept your definition. Otherwise, as I did for this simple example, click Cancel and all cases are considered.**

10. **Click the OK button and the new field, along with its counts, is generated.**

    The result is the new variable named count, as shown in Figure 7-12.

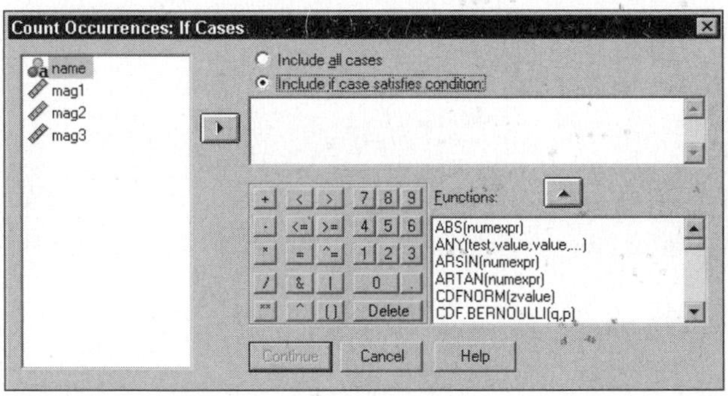

Figure 7-11: Define the criteria by which case inclusion in the count is determined.

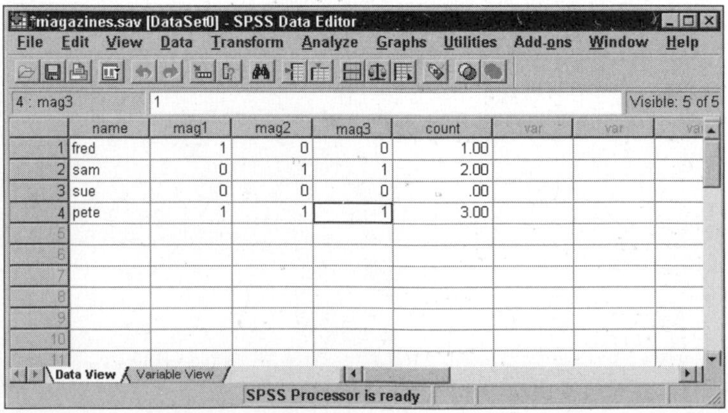

Figure 7-12: A new variable containing the total number of subscriptions per case.

Recoding Variables

You can have SPSS change specific values to other specific values according to rules you give it. You can change almost any value to anything else. For example, if you have yes and no represented by 5 and 6, you could recode the values into 1 and 2. You can recode the values in place without creating a new variable, or you can create a new variable and recode values into it. You may want to do this to correct errors or to make the data more usable.

Same-variable recoding

When you're recoding values without creating a new variable to receive the new values, be sure you've stored a safety copy of your data before you start. The changes to your data can't be automatically reversed and you could destroy information.

The following example is a list of names of individuals who were invited to an affair. If they responded with a yes, the response value was set to 1; if they responded with a no, the value was set to –1. Those with a 0 have not yet responded. As the date of the affair approaches, you decide to convert all the –1 responses to 0 to get a count of people not coming. Here's how:

To download the file, go to www.dummies.com/go/spss. You can download this simple file or all the files created for this book. Simply place the files in any directory so you can find them through the menus of SPSS.

1. Choose Open⇨File and load the rsvp.sav file.

The window shown in Figure 7-13 appears.

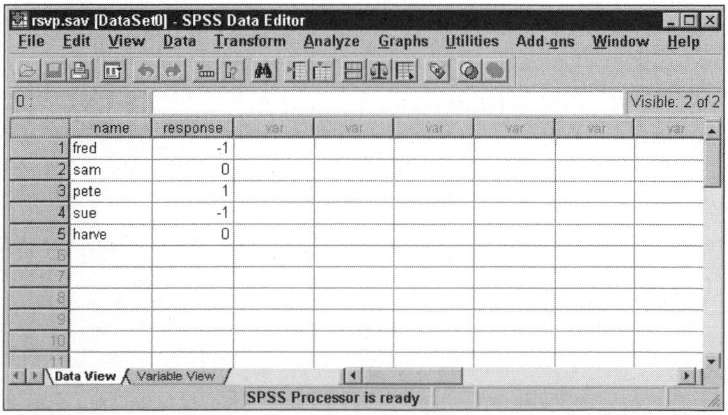

Figure 7-13: The list of names with the three possible response conditions.

2. **Choose Transform⊏>Recode into Same Variables.**

3. **Select the `response` variable and click the button with the triangle to move the variable to the panel on the right, labeled Numeric Variables, as shown in Figure 7-14.**

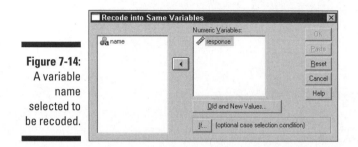

Figure 7-14:
A variable name selected to be recoded.

4. **Click the Old And New Values button.**

   The window shown in Figure 7-15 appears.

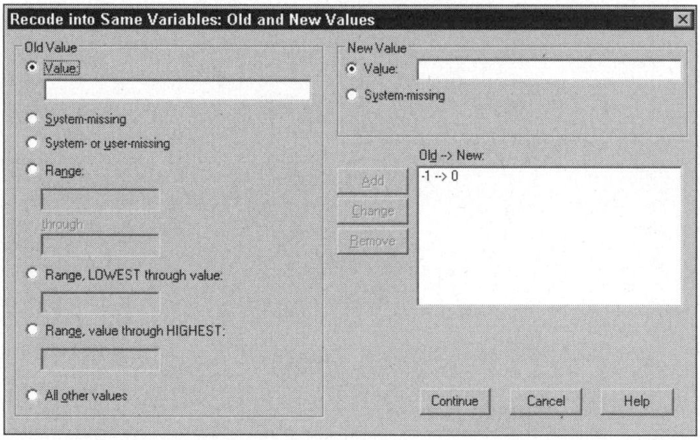

Figure 7-15:
Define the recoding of old values into new values.

5. **As shown in the figure, enter an existing value in one of the Old Value choices, and then enter a New Value for it.**

   You can specify a range of old values and have them mapped to a new value. You can also specify that the new value is to be missing and the old value will be mapped to that. You can, if you want, map a number of old values to new values and SPSS will do all the recodings at once. For each mapping of an old value to a new value, use the Add button to make the mapping appear in the window labeled Old –> New.

6. **After you have entered all the mappings (in this example it is just the one), click Continue.**

7. **Optionally, you can click the If button and the window in Figure 7-16 appears so you can limit the number of cases to which the recoding will apply.**

   The limiting is accomplished by entering an expression that must be true for a case to be included. In our example, we enter no expression, because we want the process to apply to all cases.

8. **Click the Continue button.**

   All the –1 values are converted to 0, as shown in Figure 7-17. The variable has had its values recoded.

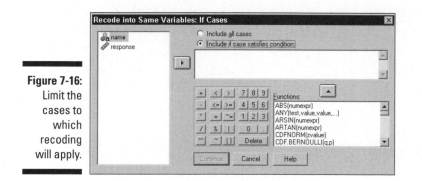

Figure 7-16: Limit the cases to which recoding will apply.

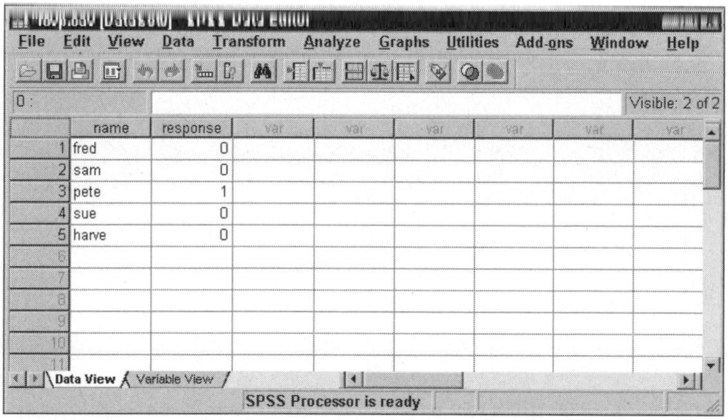

Figure 7-17: All –1 values have been recoded as 0 values.

Different variable recoding

It could be that you don't want to overwrite the existing values, but you would like to have the recoded data available. The following steps will do the same thing as in the preceding example, except the recoded values will be stored in a new variable.

1. **With the `rsvp.sav` file loaded the same as before (refer to Figure 7-13), choose Transform⇨Recode into Different Variables.**

2. **On the right, in the Output Variable area, enter a name and label for a new variable.**

   For the output variable, you can choose a new variable name and a new variable will be created, or you can choose an existing variable name and have its values overwritten.

3. **Click the Change button to move that name to the panel labeled Numeric Variable ⇨ Output Variable, as shown in Figure 7-18.**

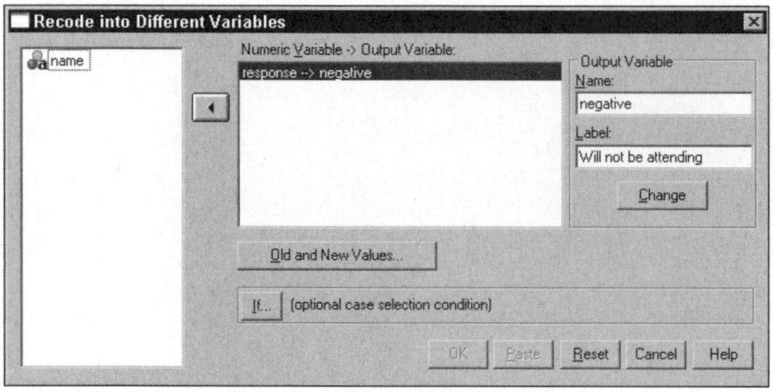

Figure 7-18:
Name the variable to receive the recoded values.

4. **Click the Old And New Values button.**

   The window in Figure 7-19 appears.

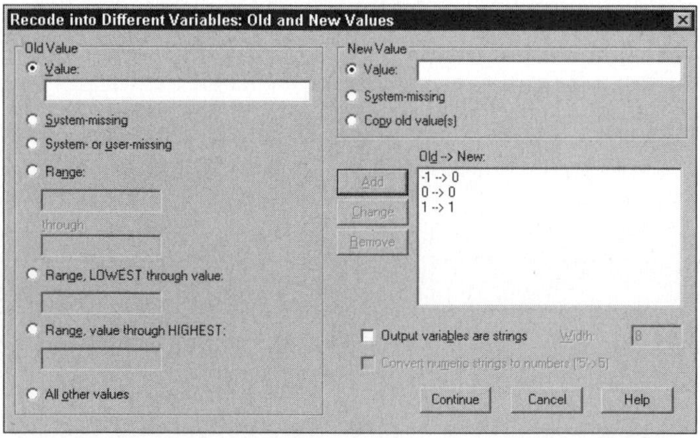

Figure 7-19:
All possible values recoded for a new variable.

5. **Define the recoding.**

   Enter an existing value into the Old Value text box and the value you want it to become in the New Value text box. Then click the Add button to add them to the Old ⇨ New list. It is important that you map all values — even the ones that don't change — because you're creating a new variable and it has no preset values.

6. **Click the Continue button.**

7. **Click the OK button.**

   The results appear, as shown in Figure 7-20. Notice that the numbers all have two digits to the right of the decimal point. This may or may not be what you want, but the new variable was created automatically and that is part of the default.

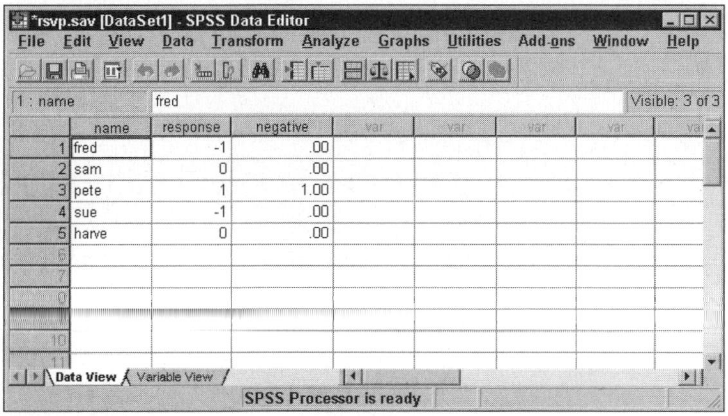

Figure 7-20:
Values
recoded
into a new
variable.

Automatic recoding

Automatic recoding converts values into something you can use in computations. For example, if you have a list of automobile names, automatic recoding will convert those names into numbers, and then you can perform an analysis on the pattern of numbers. Automatic recoding allows you to get a handle on data that could otherwise elude analysis.

To perform automatic recoding, you select options and set the names in a single dialog box. To see an example of automatic recoding in operation, follow these steps:

1. **Load `rsvp.sav` (refer to Figure 7-13).**

2. **Choose Transform⇨Automatic Recode.**

   The Automatic Recode dialog box appears.

3. **In the panel on the left, select the name of the variable you want to recode. Then click the arrow in the middle to move the variable to the panel on the right.**

4. **In the New Name text box, enter the name of the variable to receive the recoded values.**

5. **Click the Add New Name button.**

    The name you entered appears in the panel above the new name, as shown in Figure 7-21.

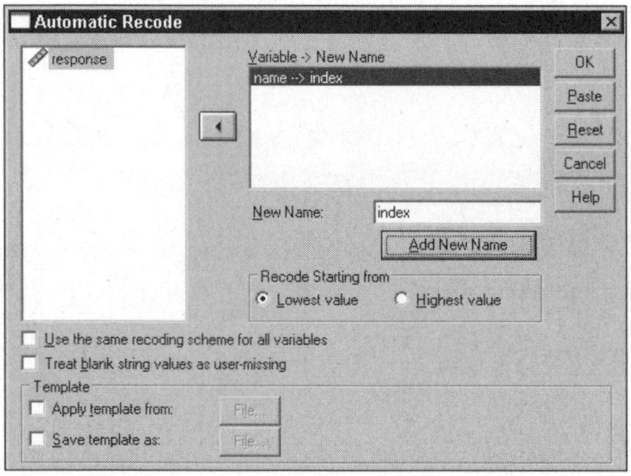

Figure 7-21: The dialog box for automatic recoding.

6. **Click the OK button and recoding takes place.**

    The result is similar to that shown in Figure 7-22, where the new variable is named index.

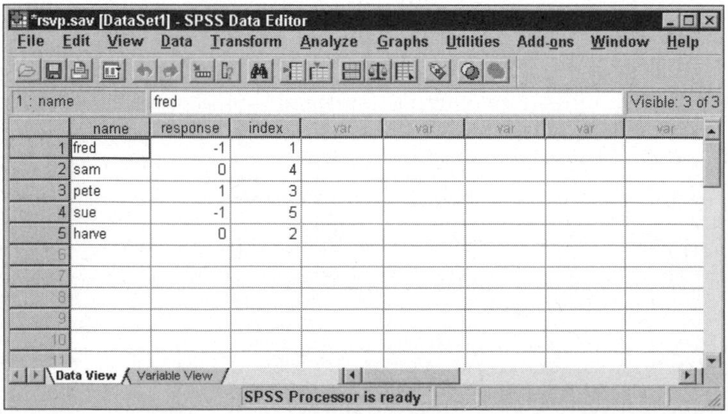

Figure 7-22: The result of automatically recoding name into index.

The values in the new variable, index, come about from sorting the values of the original variable and then assigning numbers to them in that order. If the input values are a string of characters instead of the digits of numbers, the strings are sorted alphabetically. Well, almost: Uppercase letters come before lowercase.

In the Automatic Recode window (refer to Figure 7-21), you can see the choice that allows you to recode starting the new numbers with either the lowest or the highest value. The new numeric values will be the same either way; they're just assigned in the opposite order.

At the bottom of the window are two choices for the creation of a template file. This is so you can save a file — called a *Template file* — that holds a record of the recoding patterns. That way, if you need to recode more data with the same variable names, the new input values will be compared against the previous encoding and be given appropriate values so that the two data files can be merged and the data will all fit. For example, if you have brand names or part numbers in your data, the recoding will be consistent with original values assigned the same pattern of recoded values.

Binning

If a variable is a scale variable containing a range of values, you can create groups of the values and organize them into bins. For example, you could use the ages of a number of people and put each one in its own bin — one bin for ages 0 to 20, another bin for 21 to 40, and so on. You can specify the size and content of bins in several ways. The process of actually binning is automatic.

The following steps take you through an example of the binning process by dividing salaries into bins:

1. **Choose File⇨Open⇨Data and load the salaries.sav file.**

   This file contains a list of id numbers with a salary for each, as shown in Figure 7-23.

2. **Choose Transform⇨Visual Binning.**

   The dialog box shown in Figure 7-24 appears.

3. **Select Current Salary in the panel on the left, then click the triangle in the center of the window to move the name of the variable to the panel on the right.**

4. **Click the Continue button.**

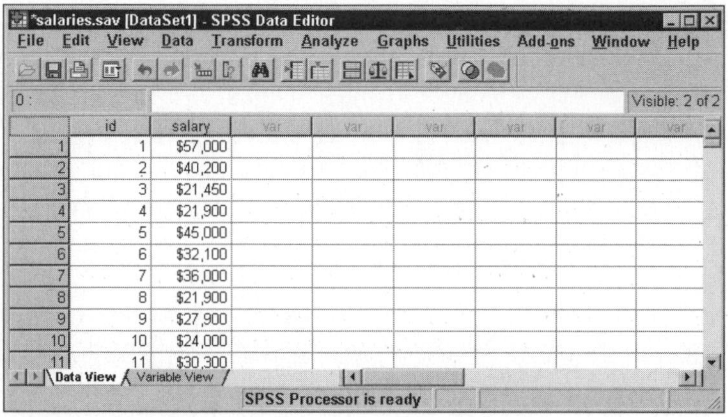

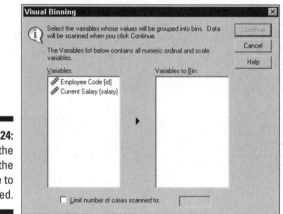

Figure 7-23:
A list of
employee id
numbers
and the
salary of
each.

Figure 7-24:
Select the
name of the
variable to
be binned.

5. **In the Scanned Variable List area, on the left, select the Current Salary label.**

   A bar graph displaying the range of values of the salaries appears in the center, as shown in Figure 7-25.

6. **Click the Make Cutpoints button.**

   A dialog box appears for specifying the size of each bin and the number of bins.

7. **Select the points at which you want to have the data cut into parts to create the bins.**

   In this example, I divided the data into even percentiles of numbers of cases — that is, each bin will contain the same number of cases, as shown in Figure 7-26. Notice that four cutpoints divide the data into five bins, each holding 20 percent of the cases. I could have chosen to divide the data into equal-width intervals — that is, each bin would contain a range of the same magnitude, which would put different numbers of cases in each bin. Also, the cutpoints could have been based on standard deviations, which would create two cutpoints, dividing the data into the three bins of low, medium, and high.

8. **Click the Apply button, and the cutpoints appear as vertical lines on the bar graph, as shown in Figure 7-27.**

   You may click the Make Cutpoints button repeatedly and cut the data different ways until you get the cutpoints the way you like. Any new cutpoints you define replace any previous ones.

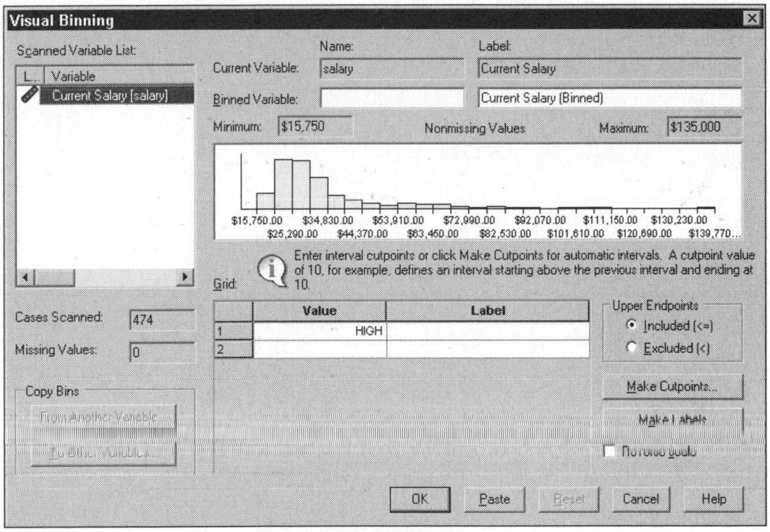

Figure 7-25: How the binning will be done.

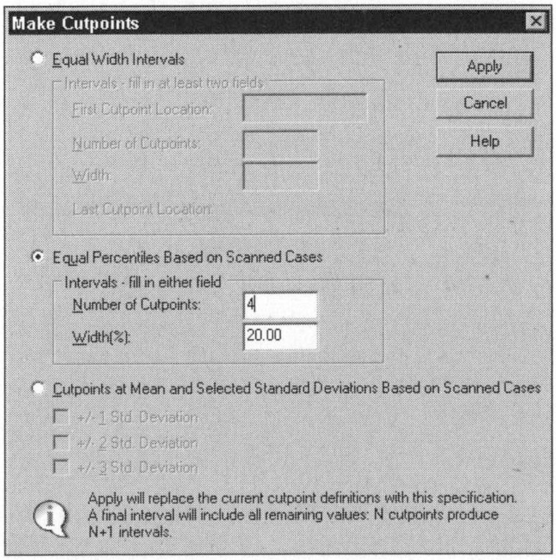

Figure 7-26: Specify how you want the data divided into bins.

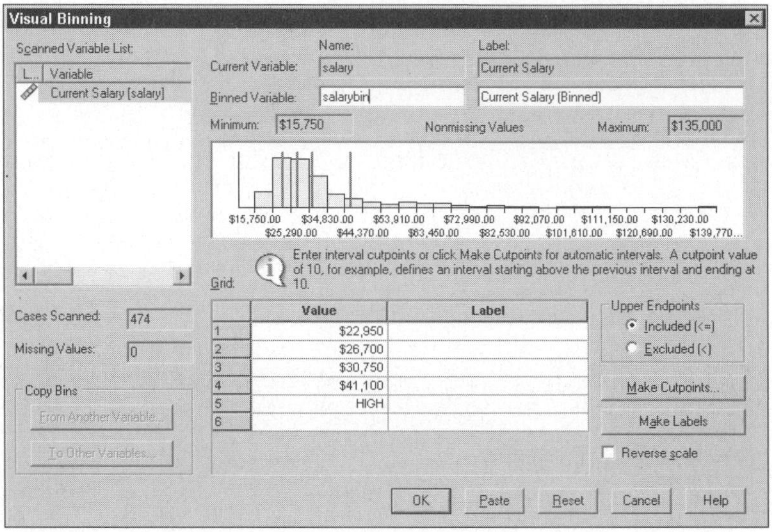

Figure 7-27:
A bar graph
of the data
with
cutpoints for
binning.

9. **Enter a name for a new variable to contain the binning information.**

   You enter the name in the Binned Variable text box. The default label for the new variable appears in the text box to the right of the name. You can change this if you want. The bins are created and numbered from 1 to 5, but if you select the Reverse Scale option in the lower-right corner, the numbering will be from 5 to 1.

10. **Click OK.**

    The new variable is created and filled with the bin values, as shown in Figure 7-28.

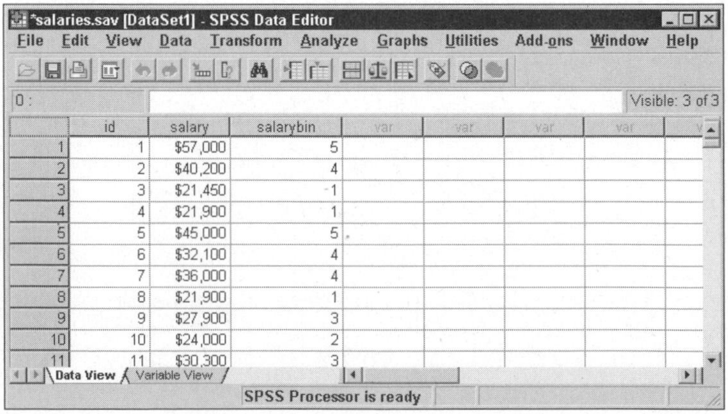

Figure 7-28:
The new
variable
containing
the bin
numbers.

The binning is now complete and you can use the new data for further analysis. One thing you can do quickly and easily is display a summary of the contents of your bins. Simply follow these steps:

1. **With the window in Figure 7-28 still on the screen, choose Transform⇨ Optimal Binning.**

2. **Select variable names on the left and click the triangular button to move the variables. Move Current Salary to Variables To Bin and move Current Salary (binned) to Optimize Bins with Respect To, as shown in Figure 7-29.**

   The variable in the Optimize Bins with Respect To text box does not have be a variable from a previous binning operation. It can be any variable that contains a collection of values sufficient for being separated into bins.

3. **Click the OK button.**

   The output is generated, as shown in Figure 7-30.

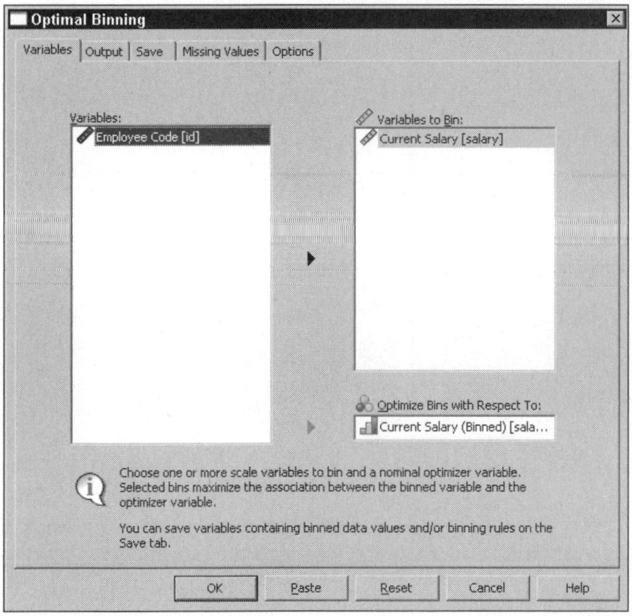

Figure 7-29: Select the bin variable and the optimizing variable.

Any variable with properly distributed values can be used as the basis of optimal binning. In the chart shown in Figure 7-30, the numbers 1 through 5 across the top are the values of the new binning variable created and stored as part of the data. The numbers 1 through 5 down the left of the graph are the result of the new binning action. The chart lets you clearly see the range of values that make up each bin.

Current Salary

Bin	End Point Lower	End Point Upper	Number of Cases by Level of Current Salary (Binned) 1	2	3	4	5	Total
1	a	$23,100	96	0	0	0	0	96
2	$23,100	$26,850	0	95	0	0	0	95
3	$26,850	$30,900	0	0	96	0	0	96
4	$30,900	$41,550	0	0	0	93	0	93
5	$41,550	a	0	0	0	0	94	94
Total			96	95	96	93	94	474

Each bin is computed as Lower <= Current Salary < Upper.

a. Unbounded

Figure 7-30: The output from optimal binning.

Chapter 8

Getting Data out of SPSS

• •

In This Chapter

▶ Outputting tables and images to the printer

▶ Outputting tables and images to the display

▶ Outputting to Excel, Word, and other applications

• •

SPSS is good at analyzing your data and displaying information you can understand in tables, charts, and graphs, but the time comes when you want to output the results to files suitable for use in other applications. You may want to send output to the printer or you may have another program that could make use of the output from SPSS. This chapter explains ways that you can output data from SPSS into forms needed by other programs.

Printing

The simplest form of output is to print the numeric rows and columns of the raw data as it appears in the Data View tab of the Data Editor window. To do this, choose File➪Print and a familiar Print dialog box appears, allowing you to select the print settings you need for your system. The table of data will be printed with lines between the rows and columns, the same as they appear on the screen. The printed form has case numbers on the left and variable names at the top.

If you're not sure what your output will look like, you can choose File➪Print Preview and see, on the screen, the same layout that will be sent to the printer. The image appears small initially, but you can click it twice and it will become as large as it will appear on paper.

If the table you're printing is too wide to fit on the sheet of paper, SPSS splits the output and places the table on multiple pages. This is done in such a way that you can hold the printed sheets side by side to get the full width of the table.

You can switch from the Data View tab to the Variable View tab and print the variable definitions. This output always requires two pages to include the full width of the table.

Exporting to a Database

You can export SPSS data directly to a database. Choose File⇨Export to Database and follow the instructions SPSS supplies for your database. SPSS knows how to write to dBase, Excel, FoxPro, Access, and text file databases. If you have a different database system, you should be able to configure SPSS for it by clicking the Add ODBS Data Source button. In similar fashion, you can read data from a database by choosing File⇨Open Database.

To export the data, simply follow the instructions on the screen regarding selecting the variables to be written and whether to write new data or overwrite existing data.

Using SPSS Viewer

Whenever you run an analysis, or produce a graph, or do anything that generates output (even loading a file), the SPSS Viewer window pops up automatically to display what you've created. This display is the most fundamental form of output from SPSS and is the first step in producing other forms of output.

Chapters 9 through 14 provide details on generating tables, graphs, and descriptive text in the SPSS Viewer window. This section describes how to output that Viewer data to files in different formats.

You can output data from SPSS Viewer in several file formats appropriate for use by other applications. Some output formats are graphics only, some are text only, and still others are a mixture of text and graphics. Some form of graphic output is usually necessary because of the graphs and charts constructed by SPSS.

In every case you begin by choosing File⇨Export, which displays the Export Output dialog box shown in Figure 8-1. In the Export drop-down list, you can choose which items in the View window to export — the entire document, the text of the document without graphics, or the graphics without text.

In the lower-left corner of the dialog box, you can select which pieces of information in SPSS Viewer you want to include as part of the output:

- ✔ All Objects outputs all the information that SPSS Viewer contains, whether or not the information is currently visible.
- ✔ All Visible Objects includes only those objects being displayed by SPSS Viewer.
- ✔ Selected Objects allows you to decide which objects to output.

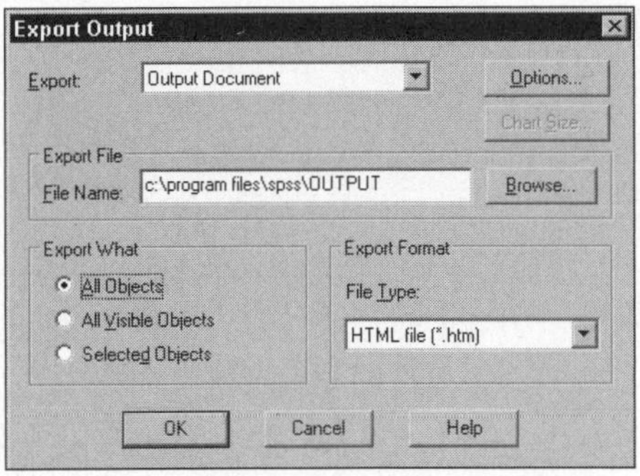

Export Output

Export: [Output Document ▼]   [Options...]

                                [Chart Size...]

Export File
File Name: [c:\program files\spss\OUTPUT]   [Browse...]

Export What   Export Format
⦿ All Objects   File Type:
○ All Visible Objects
○ Selected Objects   [HTML file (*.htm) ▼]

        [OK]   [Cancel]   [Help]

Figure 8-1:
The main
control
window for
generating
output from
SPSS
Viewer.

The set of selections made available to you in Export What is determined by the types of objects being displayed by SPSS Viewer, which ones (if any) are selected, and the choice in the Export drop-down list. The only combinations of options available are those that will produce output.

Figure 8-2 shows SPSS Viewer displaying both text and graphics. On the left is a list of names of objects. If the name of an object is visible in the list, the object itself is visible in the Viewer window. You can make objects appear and disappear by clicking the plus and minus signs. If the name of an object is highlighted in the list, the object is marked as selected in the Viewer window. A selected object appears surrounded by boxes. In the figure, the log at the top is not selected, but all other objects are. When producing output, you can select only visible objects, only selected objects, or all objects.

You can output the following types of files:

- ✔ Text file
- ✔ HTML Web page
- ✔ Excel file
- ✔ Rich Text Format (RTF), readable by Word
- ✔ PowerPoint display file
- ✔ Portable Document Format (PDF)

Some formats (for example, the text file format) require that graphics be output in separate files. You can also elect to output only graphics files. Graphics can be output in the following formats:

- ✔ Standard jpeg (JPG)
- ✔ Macintosh Pict (PCT)
- ✔ Portable Network Graphics (PNG)
- ✔ Postscript (EPS)

  ✔ Tagged Image File Format (TIFF)

  ✔ Windows Bitmap (BMP)

  ✔ Windows Metafile (WMF)

  ✔ Enhanced Metafile (EMF)

Creating an HTML Web page file

If you decide to format your output file as a Web page, the output text will be
formatted as HTML. Any pivot tables selected for output will be formatted as
HTML tables.

HTML is text only, but it can link to image files, so any graphic you select to
output will be in the image format of your choice in a separate file. You can
make a number of decisions about the details of the image file, as shown in
Figure 8-3, which appears when you click the Options button in the Export
Output dialog box.

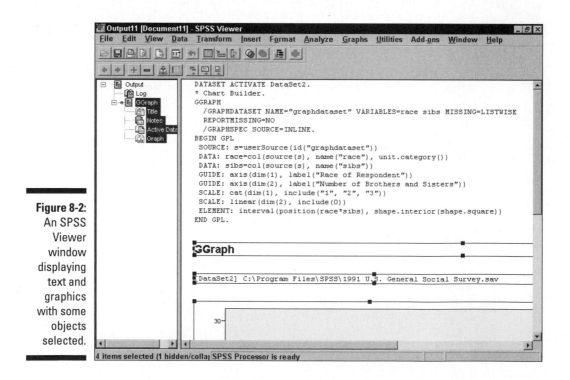

Figure 8-2:
An SPSS
Viewer
window
displaying
text and
graphics
with some
objects
selected.

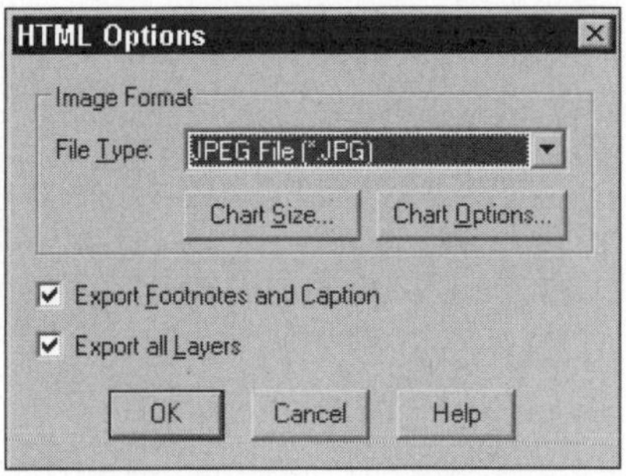

Figure 8-3:
The options
for creating
an HTML
file.

HTML Options

Image Format

File Type: JPEG File (*.JPG)

Chart Size...   Chart Options...

☑ Export Footnotes and Caption

☑ Export all Layers

OK   Cancel   Help

After you select the type of graphics file (in the figure, JPEG has been chosen), you may decide its size and other characteristics (including color type, the level of compression, and image layering). Exactly what options are available depends on the type of graphic file chosen. Unless you have a specific goal in mind, you should use the defaults. You can also choose whether to include the text of footnotes and captions as part of the HTML.

If you don't understand the bewildering options to generate your selected type of graphics file, experiment. Start with the defaults and make changes only if you need to. It doesn't cost anything, other than a few minutes of your time, to try different combinations of options and decide on the set you like.

Figure 8-4 shows part of the output page as it appears in a Web browser using the default settings for everything, including the JPEG image. Notice that the commands that generated the graphics were included and formatted in an HTML table. You may decide to leave that information out. You could, if you want, leave the table out and include only the graphic and its annotation. Also, if you were going to publish this as a Web page, you would probably want to edit the heading so that it's more descriptive and matches the style of the rest of your Web site.

In this example, the output file name is webfile, so the main file is WEBFILE.HTM and the image file is WEBFILE.0.JPG. The JPG suffix indicates that a JPEG image file was chosen. The digit in the image file name is necessary because there could be more than one and each needs a unique name.

Creating a text file

If you want to output a simple text file, you still have a number of options to choose from, as shown in Figure 8-5. The first two options are whether to use spaces or tabs to position characters on the page. This choice can be important because alignment is crucial to some data layouts, and programs might have different tab settings and change the appearance of the output when it's displayed.

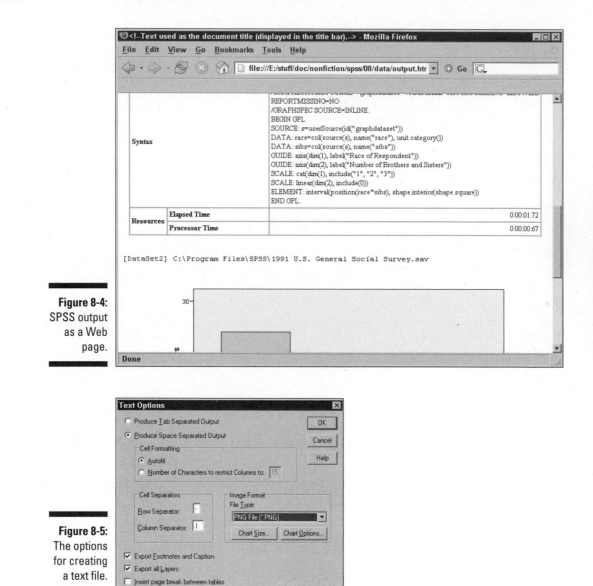

Figure 8-4:
SPSS output as a Web page.

Figure 8-5:
The options for creating a text file.

Tables output as text use certain characters to define the cells in which data items are shown. You can select any characters you want to act as separators and draw the borders, or you can accept the default of the minus sign and vertical bar, as shown in the figure. (The vertical bar is a standard keyboard character and is usually on the same key as the backward slash. It sometimes looks like a vertical line broken in the middle.) If you're outputting tables, you can choose a maximum cell size or just use the default Autofit option and let SPSS decide the number of characters that will fit in each column.

The output shown in Figure 8-6 is a simple listing in a DOS command-line window of a text file generated by SPSS. It is the same data as in the previous example, which was formatted into HTML. Also, like HTML, the graphic is output in a separate file. The text file includes the full path name of the produced graphic file. You have the same set of options for producing graphic files as you have for Web page files.

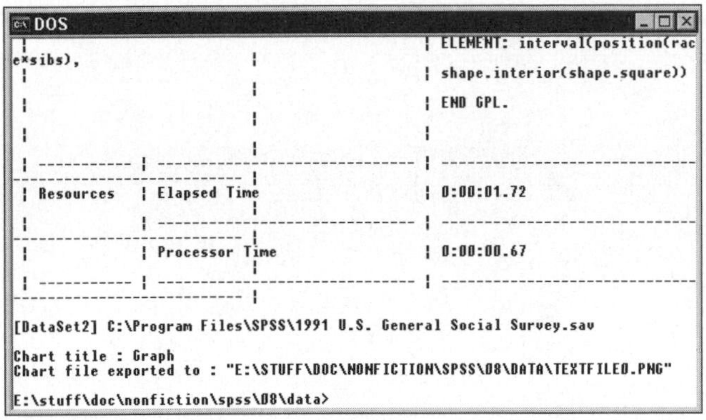

Figure 8-6:
SPSS output as a text file.

You won't use text file output unless you have an application that prefers text files as its input.

In this example, the output file name is textfile, so the main file was named TEXTFILE.TXT and the graphic file was named TEXTFILE0.PNG. The PNG suffix indicates that a PNG graphic file was chosen. The digit in the graphic file name is necessary because there could be more than one and each needs a unique name.

Creating an Excel file

Creating an Excel file is easier than creating any other kind of file because the options are so few, as shown in Figure 8-7. You get to choose whether or not the output will include footnotes and captions, and whether you want to include more than the first layer of table information. (It's possible to generate a table with one variable, such as gender, designated as a layer variable. The other variables would then display a different set of values for each sex.)

When you want to produce output, click the OK button in the Export Output dialog box, and a file is generated. The file can be loaded directly into Excel, as shown in Figure 8-8.

Figure 8-7:
The two choices you have when producing an Excel file.

Figure 8-8:
SPSS output as an Excel file.

No graphics are included in Excel files, so graphics are not output even if you choose to output everything. Graphics are ignored, but the text information that comes before and after them is included.

In this example, the output file name is `excelfile`, so the output file was named `excelfile.xls`.

Creating a Word document file

If you choose to output a Word document file, you have no graphic options to set because both text and graphics are included in one output file. The options you can choose from are shown in Figure 8-9. You choose whether to include footnotes and captions, and whether to include all layers of any tables that may be in the output.

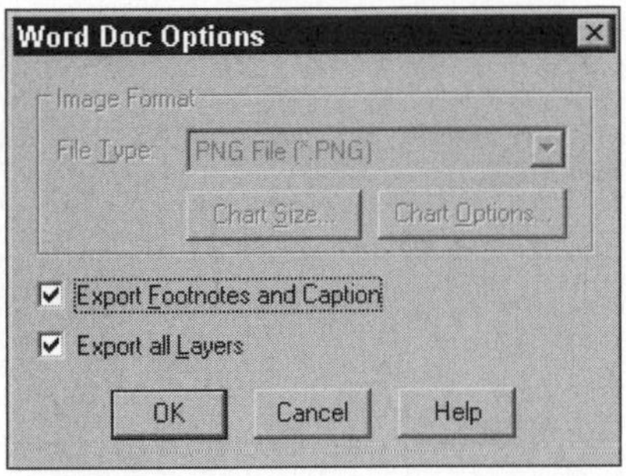

Figure 8-9:
The two choices you have in producing a Word document.

Word Doc Options

Image Format

File Type: PNG File (*.PNG)

Chart Size...   Chart Options...

☑ Export Footnotes and Caption
☑ Export all Layers

OK   Cancel   Help

When you want to produce output, click the OK button in the Export Output dialog box, and the file is generated. The output file can then be loaded directly into Word, as shown in Figure 8-10.

The output file is in RTF (Rich Text Format), a file type that can be loaded and used by most word processors, including OpenOffice, StarOffice, and WordPerfect.

In this example, the output file name is `wordfile`, so the output file was named `WORDFILE.DOC`.

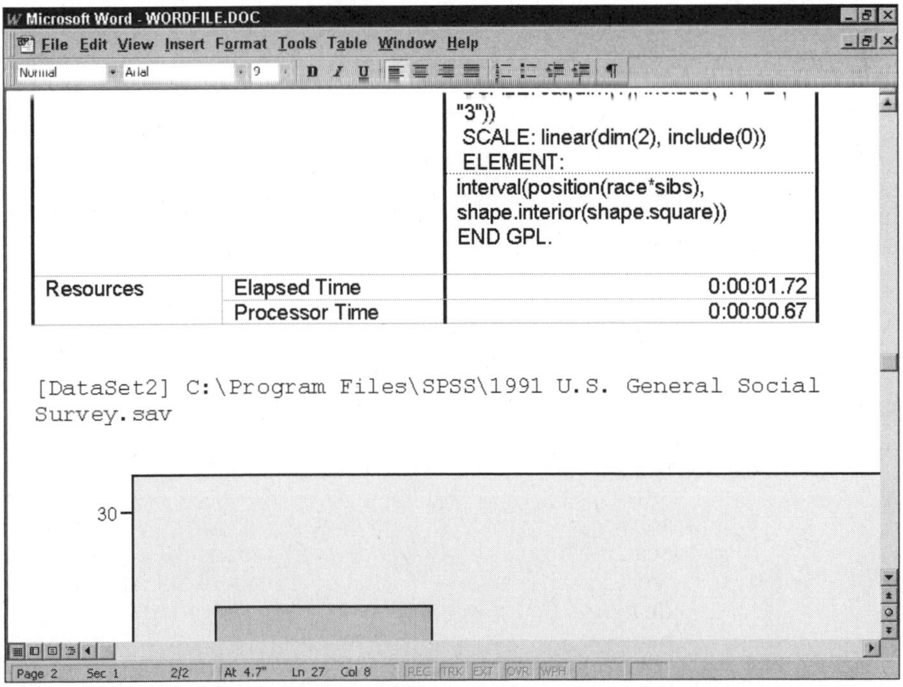

Microsoft Word - WORDFILE.DOC

File   Edit   View   Insert   Format   Tools   Table   Window   Help

Normal   Arial   9   B I U

```
                                                     "3"))
                                                      SCALE: linear(dim(2), include(0))
                                                       ELEMENT:
                                                     interval(position(race*sibs),
                                                     shape.interior(shape.square))
                                                      END GPL.
```

| Resources | Elapsed Time | 0:00:01.72 |
| | Processor Time | 0:00:00.67 |

```
[DataSet2] C:\Program Files\SPSS\1991 U.S. General Social
Survey.sav
```

30

Figure 8-10:
SPSS output as a Word/ RTF file.

Page 2   Sec 1   2/2   At 4.7"   Ln 27   Col 8   REC TRK EXT OVR WPH

Creating a PowerPoint slide document

A PowerPoint file includes only tables and graphs, so you can produce a series of display slides that contain all your graphics. The basic options are shown in Figure 8-11.

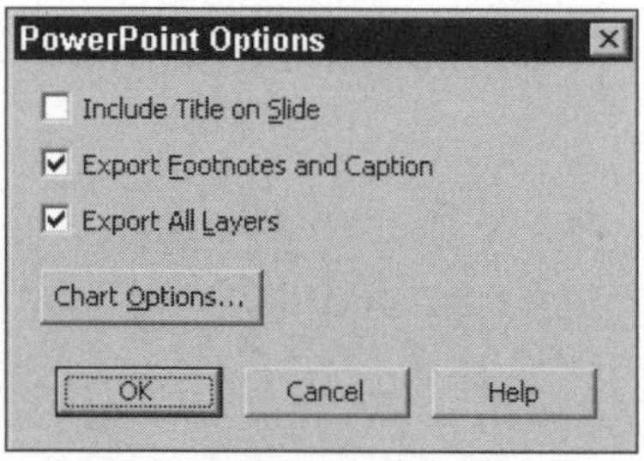

PowerPoint Options ☒

☐ Include Title on Slide

☑ Export Footnotes and Caption

☑ Export All Layers

Chart Options...

OK   Cancel   Help

You also have some options for the graphics you want to include in your slide presentation. Clicking the Chart Options button in the PowerPoint Options dialog box brings up the options for TIFF images, as shown in Figure 8-12. (All PowerPoint images are in the TIFF format.)

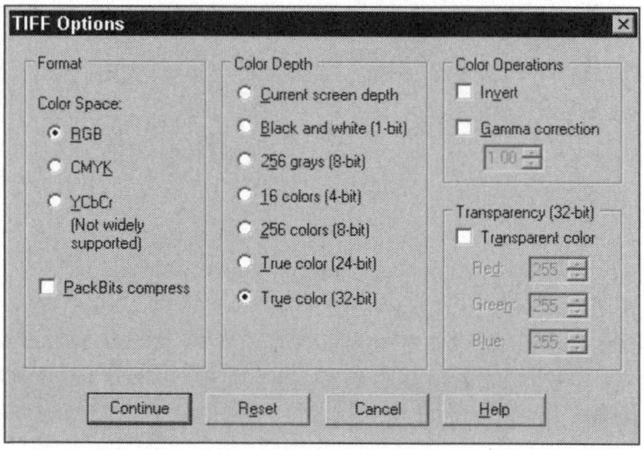

Figure 8-12:
Option
settings for
the TIFF
images
included
with
PowerPoint
slides.

TIFF Options ☒

Format

Color Space:
● RGB
○ CMYK
○ YCbCr
(Not widely supported)

☐ PackBits compress

Color Depth
○ Current screen depth
○ Black and white (1-bit)
○ 256 grays (8-bit)
○ 16 colors (4-bit)
○ 256 colors (8-bit)
○ True color (24-bit)
● True color (32-bit)

Color Operations
☐ Invert
☐ Gamma correction
1.00

Transparency (32-bit)
☐ Transparent color
Red: 255
Green: 255
Blue: 255

Continue   Reset   Cancel   Help

Your output will include only charts, graphs, and pivot tables; the rest of your data is discarded and doesn't appear anywhere in the set of produced slides. Figure 8-13 displays the slide produced from the same SPSS Viewer data that was used in the previous example. If you need some text slides before or after your graphics, you must add those yourself.

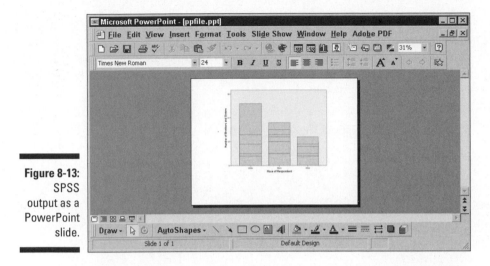

Figure 8-13:
SPSS
output as a
PowerPoint
slide.

In this example, the output file name is ppfile, so the output file was named
ppfile.ppt.

Creating a PDF document

It is becoming more common to place information on the Internet in a PDF
format instead of an HTML format. Both are read-only files, but a PDF gives
the creator of the file more control over the displayed appearance in a
viewer. An HTML page is relatively free-form when compared to a PDF file.
With a PDF file, you can put your information on the Internet and have it seen
the same way by every person who views it.

A PDF file contains formatted text and graphics, so any PDF you output
will look very much like the original data displayed in SPSS Viewer. PDF
handles graphics in a standard way, so you don't have the typical graphic
options to set. However, you do have some other options, as shown in
Figure 8-14.

The first option has to do with the action taken by the viewer when it displays your file. You can set your file to be an all-or-nothing-type file — that is,
if you don't have this check box selected, not a single page of it will be dis-
playable until the entire file has loaded. This doesn't have much effect except
for long documents, where the time it takes for them to be loaded into the

viewer is noticeable. It can be frustrating and confusing to wait for a long document to finish loading before any of it can be seen.

The second option controls the inclusion or exclusion of bookmarks. If your document is long, and the people reading the document use such things, the presence of bookmarks may be important.

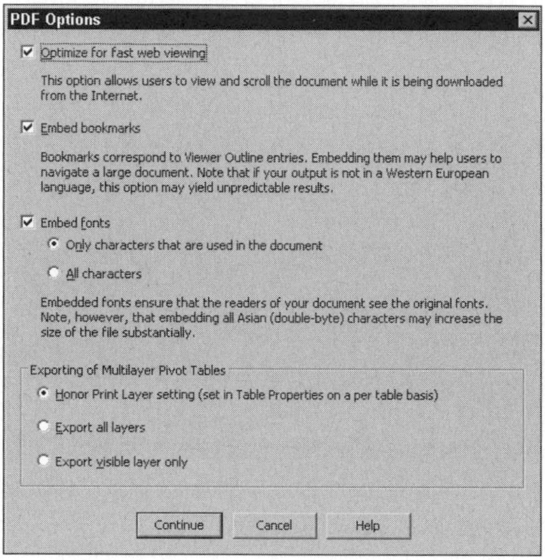

Figure 8-14:
The options
for pro-
ducing a
PDF file.

The third option has to do with the size of the document and its appearance. If the font is not important, leave it out and the PDF reader will supply one. The document it displays may not look like the original, but it will be smaller and download faster. You can include the entire font set, or you can include only the characters in your document.

The final area has to do with the layers of the pivot tables included in the for-matted document. Unless you have multilayered pivot tables in your output, this option has no effect. If you have multilayered pivot tables, you probably want to experiment to see which way you like it.

Using the default settings, SPSS produced the PDF file shown in the Adobe Acrobat Viewer in Figure 8-15.

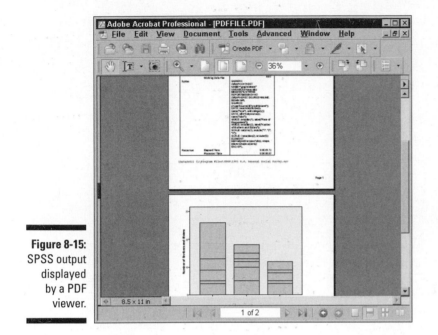

Figure 8-15:
SPSS output
displayed
by a PDF
viewer.

Part III
Graphing Data

The 5th Wave   By Rich Tennant

PROFIT
INCREASE
STRATEGY

"Look-what if we just increase the size of the charts?"

In this part . . .

Data often makes more sense if it's displayed as a graph rather than as columns or tables of numbers. You will find lots of different kinds of graphs, and some are more suitable than others for displaying your data.

SPSS makes it easy to display data in different graphical formats so you can choose the one you like. You do the clicking and SPSS does the formatting.

Chapter 9

Fundamentals of Graphing

- -

In This Chapter

▶ Building graphs easily

▶ Building graphs quickly

▶ Building graphs the old way

▶ Editing a graph

- -

*O*ver the years, the SPSS software has improved its methods for generating graphic displays of data. You can take the easy way and be guided through every step, or you can take a faster way and simply set the options to build the graph you want. The older methods of producing graphic output are still available and on the menu, so if you like to suffer while you work, you can use the procedures developed in previous years. In any case, you never have to worry about the size of text and graphics and you don't have to think about the placement of the graph on the page — SPSS does all the grunt work for you.

SPSS can display your data in a bar chart, a line graph, an area graph, a pie chart, a scatterplot, a histogram, a collection of high-low indicators, a box plot, and a dual-axis graph. Adding to the flexibility, each of these basic forms can have different appearances. For example, a bar chart can be two- or three-dimensional, in different colors, and with simple lines or I-beams for bars. The choice of layouts is almost endless.

In the world of SPSS, the terms *chart* and *graph* mean the same thing and are used interchangeably.

The Graphs menu in the SPSS Data Editor window has four options. The first three — Chart Builder, Interactive, and Legacy Dialogs — are different ways of doing the same job. (The fourth menu selection, Map, is for doing a different job.) Choosing Legacy Dialogs allows you to build graphs the original way. A better way of building graphs was devised a few years later and named Interactive — that's when the original way of building graphs became known as Legacy. Later yet, an even better procedure for building charts was devised and was added to the menu as Chart Builder. All three building methods are in place primarily for people who are in the habit of using the older procedures, but if you build a lot of graphs, you may find advantages and uses for all of them. You can get the same graphs from all three; only the process is different.

Building Graphs the Easy Way

SPSS contains Chart Builder, which uses a graphic display to guide you through the steps of constructing your display. It checks what you're doing as you proceed and won't allow you to try to use things that won't work. If the OK button is available for clicking after you've defined what you want as a result, a chart will be produced.

Gallery tab

The following example steps you through the process of creating a bar chart from example data, but you can use the same procedure to build a chart of any design. Follow this tutorial once to see how it all works. Later, you can use your own data and choices.

You can't hurt your data by generating a graphic display. Even if you thoroughly mess up the graph, you can always redo it without fear. This is one place where mistakes don't cost anything. And nobody's watching.

The following steps build a bar chart:

1. **Choose File⇨Open⇨Data and load the `1991 U.S. General Social Survey.sav` file, which is in the SPSS directory.**

2. **Choose Graphs⇨Chart Builder.**

   The Chart Builder dialog box appears, as shown in Figure 9-1. If a graph was generated previously, the display will be different; click the Reset button to clear the Chart Builder display.

3. **Make certain the Gallery tab is selected.**

4. **In the Choose From list, select Bar as the graph type.**

   The fundamental types of bar charts appear in the gallery to the right of the list.

5. **Define the general shape of the bar graph to be drawn.**

   You can do so in two ways. The simplest is to choose one member from the set of diagrams of bar graphs appearing immediately to the right of the list. You select one in the upper-left corner and drag it to the large panel at the top. Alternatively, you can click the Basic Elements tab (instead of the Gallery tab) and drag one image from each of the two displayed panels to the panel on top, thus constructing the same diagram as the bar graph. Figure 9-2 shows the appearance of the window after the dragging is complete. The result is the same no matter which procedure you follow.

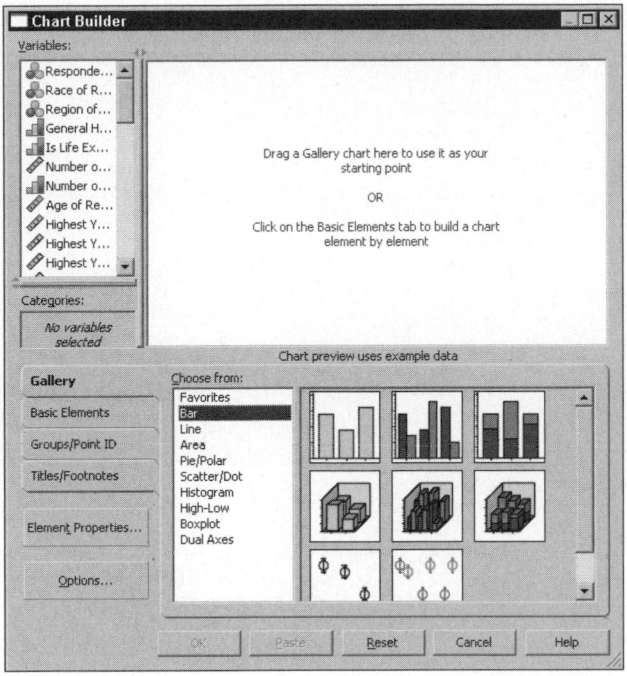

Figure 9-1:
The initial
Chart
Builder
window
with Bar
chosen.

You can always back up and start over. Anytime during the design of a graph, click the Reset button. Anything you dragged to the display panel will be deleted, and you can start from scratch.

6. **Click Close to close the Element Properties window (see Figure 9-3).**

   This window should have popped up when you dragged the graphic layout to the panel. This dialog box is not needed for this example, so you can close it. If it didn't appear but you'd like to see it, you can click the Element Properties button any time.

7. **From the list on the left, select the variable with the label and name Highest Year of School Completed (Educ) and drag it to the Y-Axis label in the diagram.**

8. **In similar fashion, select the variable with the label and name Region of the United States (region) and drag it to the X-Axis label in the diagram.**

   The screen now looks like the one shown in Figure 9-4.

   The graphics display inside the Chart Builder window *never* represents your actual data, even after you insert variable names. All that is displayed is a diagram that demonstrates the composition and appearance of the graph that will be produced.

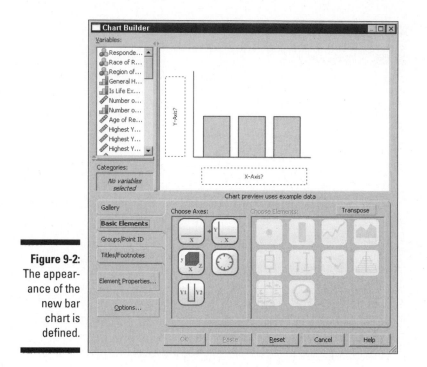

Figure 9-2:
The appear-
ance of the
new bar
chart is
defined.

9. **Click the OK button to produce the graph.**

   An SPSS Viewer window appears, containing the graph shown in
   Figure 9-5. This graph is based on the actual data and shows that the
   average number of years of education varied little from one part of
   the country to the next in this survey.

These steps demonstrate the simplest way possible of generating a chart.
Most of the options available to you were left out of the example so it would
demonstrate the simplicity of the basic process. The following sections
describe the options.

Basic Elements tab

The example in the preceding section used the Gallery tab to select the type
and appearance of the chart. Alternatively, you can click the Basic Elements
tab in the Chart Builder dialog box and select one part of the chart from each
of the two panels shown in Figure 9-6.

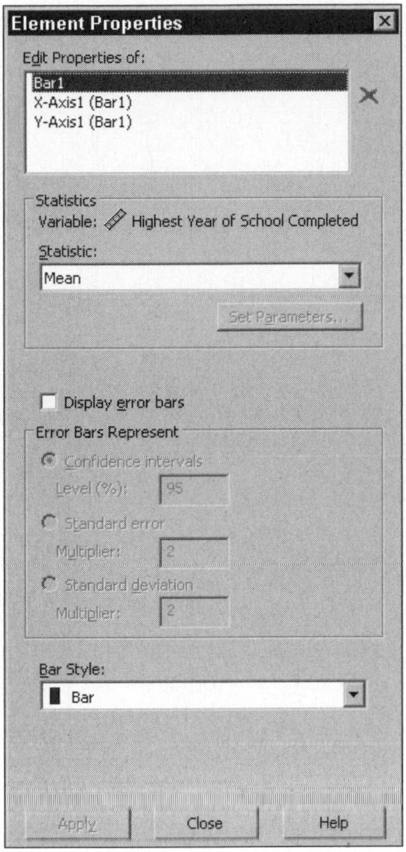

Element Properties
Edit Properties of:
Bar1
X-Axis1 (Bar1)
Y-Axis1 (Bar1)
Statistics
Variable: Highest Year of School Completed
Statistic:
Mean
Set Parameters...
Display error bars
Error Bars Represent
Confidence intervals
Level (%): 95
Standard error
Multiplier: 2
Standard deviation
Multiplier: 2
Bar Style:
Bar
Apply
Close
Help

Figure 9-3: Use the Element Properties window to modify chart elements.

Mean Highest Year of School C...
North East
South East
West
Region of the United States

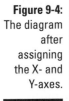

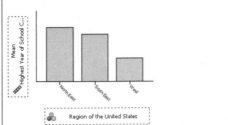

Figure 9-4: The diagram after assigning the X- and Y-axes.

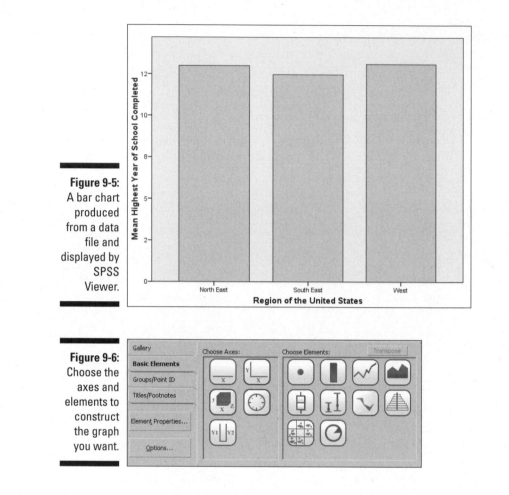

Figure 9-5:
A bar chart produced from a data file and displayed by SPSS Viewer.

Figure 9-6:
Choose the axes and elements to construct the graph you want.

It's sort of like the menu in a Chinese restaurant — you choose one from column A and another from column B. You drag one image from each panel into the panel at the top, and they combine to construct a diagram of the graph you want.

The result is the same as you get by using the Gallery tab. However, by using the Basic Elements tab, you build the graph from its components. Whether you use this technique or the Gallery depends on your conception of the graph you want to produce.

Groups/Point ID tab

Clicking the Groups/Point ID tab in the Chart Builder dialog box provides you with a group of options you can use to add another dimension to your graph.

In the example in Figure 9-7, I selected the Rows Panel Variable option, which generates a family of graphs. The new dimension adds a separate graph for the number of children in the family. A separate set of bars is drawn for those with no children, another set for those with one child, another for those with two children, and so on.

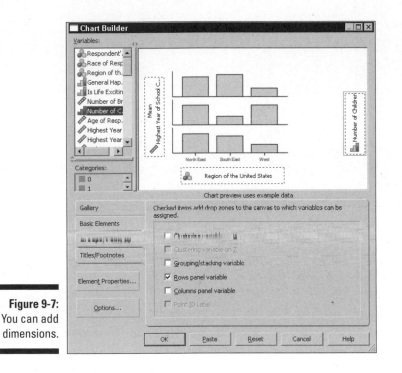

Figure 9-7:
You can add dimensions.

The Columns Panel Variable option enables you to add a variable along the other axis, thus adding another dimension. Adding variables and new dimensions this way is known as *paneling,* or *faceting.*

Clustering (gathering data into groups) can also be done along the X- or Y-axis if the variables are the type that will cluster (or bin) properly.

Titles and footnotes tab

Figure 9-8 shows the window you get when you click the Titles/Footnotes tab in the Chart Builder dialog box. Each option in the bottom panel places text at different locations on the graph. When you select an option, the Element Properties window appears so you can enter the text for the specified location.

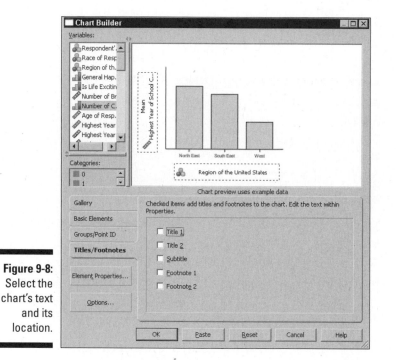

Figure 9-8: Select the chart's text and its location.

Element Properties dialog box

You can use the Element Properties dialog box at any time during the design of a chart to set the properties of the individual elements in the chart. The dialog box, shown in one mode in Figure 9-3 and another in Figure 9-9, changes every time you choose a different member from the list at its top.

The dialog box often pops up on its own when you add an item to the graph's definition, but you can make it appear any time you want by clicking the Element Properties button in the Chart Builder dialog box.

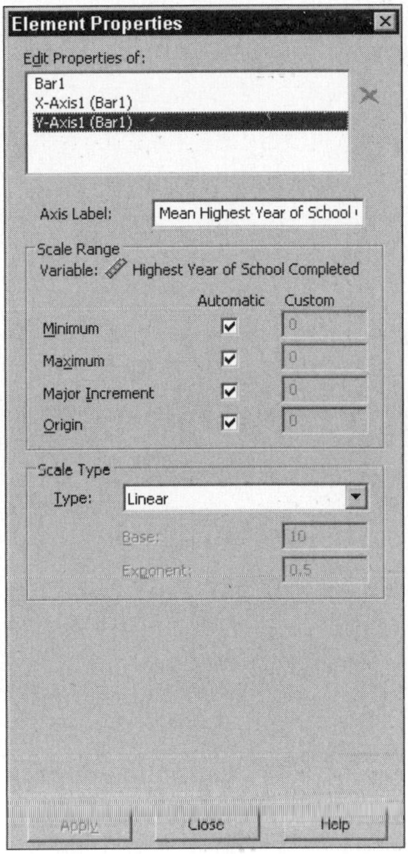

Figure 9-9:
The options
for an axis
variable.

Okay, the upcoming list of options is long, but four facts make them simple to use:

- ✔ All options have reasonable defaults. You don't have to change any of them unless you want to.

- ✔ You can always back up and change whatever settings you made. Nothing is permanent, so you can make changes until you've finished or run out of time and decide, "That's good enough."

- ✔ Not all options appear at once. Only a few show up at a time. In fact, you'll probably never see some options.

- ✔ All options become obvious once you see what they do. You don't have to memorize any of them, but you'll find they are easy to remember.

The following is a simple explanation of all the options. If an Element Properties dialog box pops up while you're building a graph, you can look up its contents in this list:

- **Edit Properties Of:** This list, which appears at the top of the window, is used for selecting which element in the chart you want to edit. Each element has a type, and the type of the element you select determines the other options available in the window. The selected element is highlighted in the diagram of the graph in some way.

- **X:** When an element is selected and the X button to the right of the list becomes enabled, clicking the button will remove the element from the list and from the graph.

- **Arrow:** For charts with dual Y-axis variables, the arrow to its right in the list indicates which of the variables will be drawn on top of the other. You can click the arrows to change the drawing order.

- **Statistic:** For certain elements, you can select the statistic (the type of value) they will display. For example, you can select Count and use simple numeric values. You can also select Sum, Median, Variance, Percentile, or any one of as many as 32 statistic types. Not all types of charts have that many options, and which options are available also depends on the types of variables displayed. For certain statistics options — such as Number in Range and Percentage Less Than — the Set Parameters button is activated and you need to click it to set the parameters controlling your choice.

- **Axis Label:** You can change the text used to describe a variable. By default, the variable label is used.

- **Automatic:** The range of the selected axis is determined automatically to include all the values of the variable being displayed along that axis. This is the default.

- **Minimum/Maximum:** You can override the Automatic default and choose the extreme values that determine the start and end points of an axis.

- **Origin:** Specifies a point from which chart information is graphed. This option has different effects for different types of charts. For example, choosing an origin value for a bar chart can cause bars to extend both up and down from a center line.

- **Major Increment:** The spacing that determines placing tick marks along with numeric or textual labels on an axis. The value of this option determines the interval of spacing when you also specify the minimum and maximum values.

- **Scale Type:** You have four choices for the scaling of an axis:

  - **Linear:** A simple, untransformed scale. This is the default.

  - **Logarithmic (standard):** Transforms the values into logarithmic values for display. You can also select a base for the logs.

- **Logarithmic (safe):** Same as standard logarithms, except the formulas that calculate values can handle 0 and negative numbers.

- **Power:** Raises the values to an exponential power. You can select an exponent other than the default value of 0.5 (which is the square root).

✔ **Sort By:** You can select which characteristic of a variable will be used as the sort key. It can be one of the following three:

- **Label:** For a nominal variable, sorting is by the names assigned to the values. This selection can be set to sort in ascending or descending order.

- **Value:** Uses the numeric values for sorting. This selection can be set to sort in ascending or descending order.

- **Custom:** Uses the order specified in the Order List.

✔ **Order List:** The list of possible values is flanked by up and down arrows. The sorting order is changed by selecting a value and clicking an arrow to move the selection up or down. To remove a value from the produced chart, select its name in the list and click the X button; the value is moved to the Excluded list. Making a change to the Order List automatically switches Sort By to Custom.

✔ **Excluded:** Any value excluded from the Order List appears in this list. To move a value back to the Order List (which also causes the value to reappear on the chart), select its name and click the arrow to the right of the list.

If a value or a margin annotation representing a value is unexpectedly missing from the graphic produced from your selections, look in the Excluded list. You may have excluded too much.

✔ **Collapse:** If you have a number of values that seldom occur, you can select this option to have them gathered into an "other" category. You specify the percentage of the total number of occurrences to make it an "other" value.

✔ **Error Bars:** For Mean, Median, Count, and Percentage, confidence intervals are displayed. For Mean, you must choose whether the error bars will represent the confidence interval, a multiple of the standard error, or a multiple of the standard deviation.

✔ **Bar style:** You can choose one of three possible appearances of the bars on a bar graph.

✔ **Categories:** You can choose how the values will be ordered when they are placed along an axis. You can select ascending or descending order. If the variable is nominal, you can select the individual order and even specify values to be left out.

✔ **Small/Empty Categories:** You can choose to include or exclude missing value information.

✔ **Display Normal Curve:** For a histogram, you can choose to have a normal curve superimposed over the chart. The curve will use the same mean and standard deviation values as the histogram.

✔ **Stack Identical Values:** For a chart that will display as points, you can choose whether points at the same location should appear next to one another or one on top of the other (blotting out the one below).

✔ **Display Vertical Drop Lines between Points:** For a chart that will display as points, any points with the same X-axis values will have a vertical line drawn joining them.

✔ **Plot Shape:** For a dot plot, you can choose

- **Asymmetric:** Stacks the points on the X-axis. This is the default.

- **Symmetric:** Stacks the points centered around a line drawn horizontally across the center of the screen.

- **Flat:** The same as Symmetric except no line is drawn.

✔ **Interpolation:** For line and area charts, the algorithm used to calculate how the line should be drawn between points:

- **Straight:** Draws a line directly from one point to the next.

- **Step:** Draws a horizontal line through each point, and the ends of the horizontal lines are connected with vertical lines.

- **Jump:** Draws a horizontal line through each point, but the ends of the lines are not connected.

- **Location:** For Step and Jump interpolation, this option causes the actual point to be indicated.

- **Interpolation through Missing Values:** For Straight, Step, or Jump, this option draws lines through missing values. Otherwise, the line shows a gap.

✔ **Anchor Bin:** The starting value of the first bin. This option is available for histograms.

✔ **Bin Sizes:** The sizes of the bins when producing a histogram.

✔ **Angle:** Rotates a pie chart by selecting the clock position at which the first value starts. You can also specify whether the values should be included clockwise or counterclockwise.

✔ **Display Axis:** For a pie chart, you can choose to display the axis points on the outer rim.

Options

Clicking the Options button in the Chart Builder dialog box opens the Options dialog box, shown in Figure 9-10.

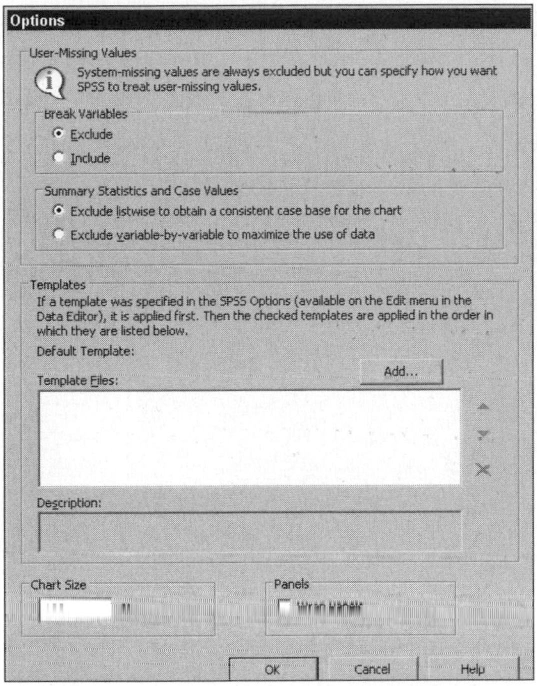

Options

User-Missing Values
ⓘ System-missing values are always excluded but you can specify how you want SPSS to treat user-missing values.

Break Variables
◉ Exclude
○ Include

Summary Statistics and Case Values
◉ Exclude listwise to obtain a consistent case base for the chart
○ Exclude variable-by-variable to maximize the use of data

Templates
If a template was specified in the SPSS Options (available on the Edit menu in the Data Editor), it is applied first. Then the checked templates are applied in the order in which they are listed below.
Default Template:
Template Files: Add...

Description:

Chart Size   Panels
                                   ☐ Wrap panels

OK   Cancel   Help

Figure 9-10: Options you can apply to a chart.

When you define the characteristics of a variable, you can specify that certain values be considered missing values. The options in the Break Variables area let you decide whether you would like those included or excluded from your chart. You can also specify how you would like summary statistics handled. Missing values are discussed in Chapter 4, and the different types of summary data are described in Chapter 7.

Templates are files that contain all or part of a chart definition. You can insert one or more template file names into the list in this window and have the definitions applied as the default starting point for all charts you build. You create a template file from a finished chart displayed in SPSS Viewer. You find out more on making templates later in this chapter.

Templates only come in handy when you need to build lots of similar charts.

You can use the Chart Size option to make the generated charts smaller or larger.

The Wrap Panels option determines how the panels are displayed when you have a number of them in a chart. SPSS is using the word *panel* to refer to the rectangular area in SPSS Viewer in which a chart is placed. Normally, the panels are shrunk to fit, but if you select this option they remain full size and wrap to the next line.

Building Graphs the Fast Way

The charts you build by choosing Graphs⇨Interactive are the same you build by using Chart Builder, but you get less guidance along the way. I suggest not doing it this way until you are familiar with Chart Builder. Although the Interactive option is much faster — you just make selections and go — there is no diagram of the chart to remind you of where you are and what you've accomplished.

The first thing you select is the kind of chart you want to build (such as bar, dot, or line). This takes you to a window filled with options, like the one shown in Figure 9-11 for bar graphs. Notice the tabs along the top of the window. You click those to change the options. Once you have the options set, click the OK button and the chart is generated.

The Reset button removes everything you've entered in all the tabbed windows and restores all the defaults.

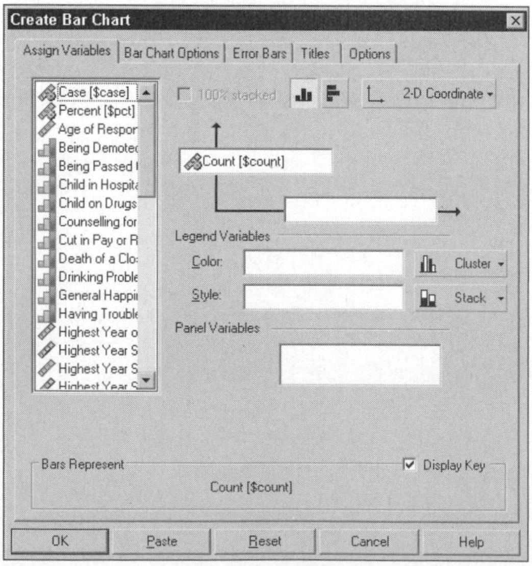

Figure 9-11:
The Interactive options for constructing a bar graph.

The Help button provides some information about whichever list of options is displayed at the moment.

The Paste button is for those who want to add to their graph definition in the most fundamental way possible. A graph is actually constructed by a command in the Syntax Command language. The steps you took to create a graph actually created the Command Syntax, which in turn was used to create the graph. The Paste button opens SPSS Syntax Editor with the Command Syntax in it, so you can edit the text of the command to produce the chart the way you want. For more on using the Syntax language, see Chapters 15 and 16.

Building Graphs the Old-Fashioned Way

The charts you build by choosing Graphs⇨Legacy are simpler forms of the ones you build in the other ways, and the process is a bit different. As in the Interactive process, you don't have the graphics and guidance you get from Chart Builder. The windows you use to set the variables are different from those of the Interactive approach. You don't have as many decisions to make, but you still need to be familiar with the process.

The first selection you make is the type of chart to be produced (such as bar, dot, or line). As you proceed through the steps of the definition of the graph, different windows appear, like the one shown for bar graphs in Figure 9-12. Each time you finish with one window, you click the OK button and move to the next window in the series. When you finish the last one, the result appears.

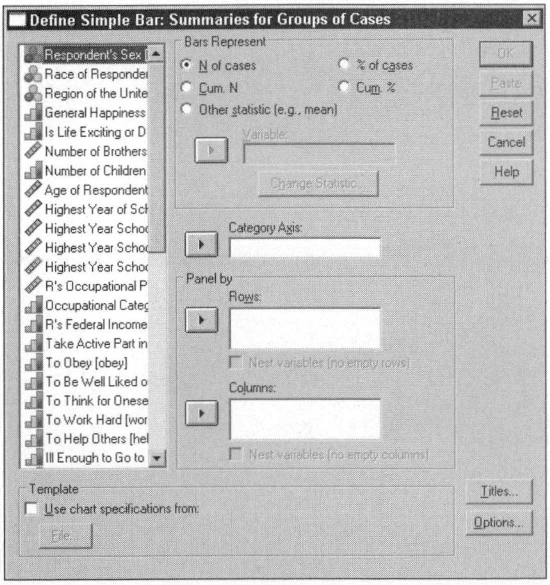

Figure 9-12:
A window for designing a bar chart using the Legacy method.

The options presented to you in the Legacy method are not quite as complete as the ones in the other two methods. This makes it easy to produce simpler charts and graphs, but you must know what you're doing because you can't back up. Once you've decided on the values in a window and move on, the values stay that way until you've finished.

Editing a Graph

After you've built a chart and it's displayed in SPSS Viewer, you can still change it. Double-click the graph, and a copy of it appears in a new Chart Editor window, as shown in Figure 9-13.

Earlier in this chapter I mentioned that you could use templates to help define new charts in Chart Builder. You can create a template file from Chart Editor by choosing File⇨Save Chart as Template and entering a file name.

Using Chart Editor, you can do a number of things with the chart. The options available are mostly the same ones you worked with when defining the original layout, so there are no surprises.

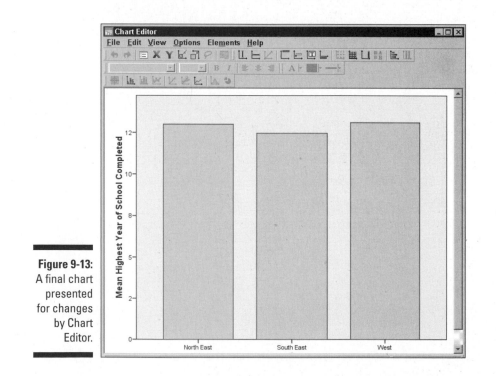

Figure 9-13:
A final chart presented for changes by Chart Editor.

Figure 9-14 is the same graph as Figure 9-13, but with the axes transposed (to make the bars grow horizontally), grid lines displayed (to mark the relative extents of the bars), the overall size of the chart reduced, and the value of the height of each bar displayed in its middle.

The many menus of Chart Editor have option settings that you can use to try to make your chart demonstrate the data. None of the selections are destructive — if you try something and don't like it, back your changes out and restore what you had before.

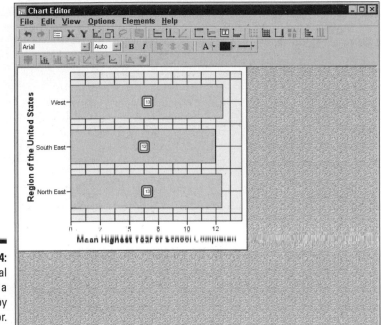

Figure 9-14:
The final chart after a few edits by Chart Editor.

Chapter 10

Some Types of Graphs

In This Chapter

▶ Drawing line charts with single and multiple lines

▶ Generating scatterplots

▶ Creating bar charts from your data

This chapter, and the next one, contain examples of different kinds of graphic displays of data. This chapter shows you how to build the ones you're probably most familiar with. Each example is presented as a step-by-step procedure, and each example is kept as simple as possible.

These two chapters don't present every variation of every possible chart, but you can certainly use the procedures presented here to produce some nifty-looking graphs. And once you get the basic idea of producing graphs, you should have no problem branching out and making fancy graphs of your own.

You could work through the examples in these two chapters to get an overview of building the kind of graphs you can get from SPSS — not a bad idea for a beginner — or you could just choose the look you would like your data to have and find out how to construct the chart by stepping through the example that produces it. Either way, after you clearly understand the basics, you can step through the process again and again, with variations, using your data to make your charts appear the way you would like them to.

Line Chart

A *line chart* works well as a visual summary of categorical values. Line charts are also useful for displaying a timeline because they demonstrate up and down trends so well. Line graphs are popular because they are easy to read. If they're not the most common type of statistical chart, they're a contender for the title.

The display of data is similar in a line chart and a bar chart. If you decide to display data as a line graph, you should probably try the same data as a bar chart to see which you prefer.

Simple line chart

The following steps generate a simple line chart displaying a single timeline:

1. **Choose File⇨Open⇨Data and open the `Employee data.sav` file, which is in the SPSS installation directory.**

2. **Choose Graphs⇨Chart Builder.**

   The Chart Builder dialog box appears.

3. **In the Choose From list, select Line.**

4. **Drag the diagram on the left (the one with the single line) to the panel at the top.**

5. **In the Variables list, drag Current Salary to the Y-Axis rectangle in the panel at the top.**

6. **Again in the Variables list, drag Date of Birth to the X-Axis rectangle in the panel.**

7. **Click the OK button.**

   The chart in Figure 10-1 appears.

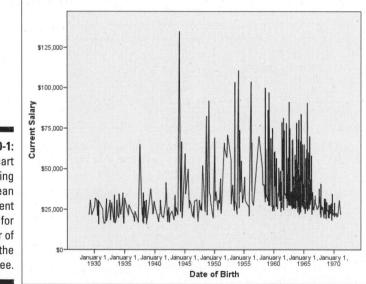

Figure 10-1:
A line chart displaying the mean current salary for the year of birth of the employee.

A chart with multiple lines

You can have more than one line appear on a chart by adding more than one variable name to an axis. But the variables must contain a similar range of values before they can be represented by the same axis. For example, if one variable ranges from 0 to 1000 pounds and another variable ranges from 1 to 2 pounds, the values of the second variable will show up as a straight line regardless of how they actually fluctuate.

The following steps generate a multiline graph:

1. **Choose File⇨Open⇨Data and open the `Cars.sav` file.**

   The file is in the SPSS installation directory.

2. **Choose Graphs⇨Chart Builder.**

3. **In the Choose From list, select Line to specify the general type of graph to be constructed.**

4. **To specify that this graph should contain multiple lines, select the diagram on the right (the one containing multiple lines on the displayed diagram) and drag it to the panel at the top.**

5. **In the Variables list, select Engine Displacement and drag it to the Y-Axis rectangle in the panel at the top.**

   The word *Mean* is added to the annotation because the values displayed on this axis will be the mean values of the engine displacement.

6. **In the Variables list, select Horsepower and drag it to the Y-Axis also.**

   Be careful how you drop Horsepower. To add Horsepower as a new variable, you want to drop it on the little box containing the plus sign, as shown in Figure 10-2. If you drop the new name on top of the one that's already there, the original variable is replaced.

7. **When the Create Summary Group window appears, telling you that SPSS is combining the two variables along the Y-Axis, click the OK button.**

8. **In the Variables list, select Number of Cylinders and drag it to the rectangle named X-Axis in the diagram.**

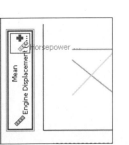

Figure 10-2:
Adding
another
variable to
the Y-axis.

9. **Click the OK button.**

The chart shown in Figure 10-3 appears.

The variables you choose as members of the Y-axis must have a similar range of values to make sense. For example, if you were to choose age and annual income as two variables to be charted together, the result would not be interesting because the salary values are in the thousands and the ages would all appear in a single line.

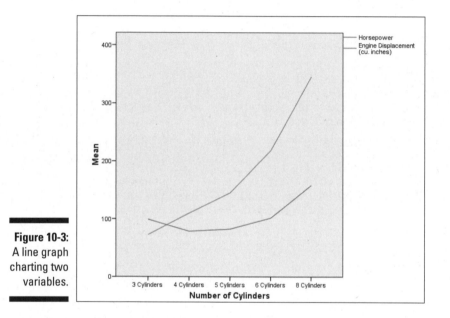

Figure 10-3:
A line graph
charting two
variables.

Scatterplots

A *scatterplot* is simply an X-Y plot where you don't care about interpolating the values — that is, the points are not joined with lines. Instead, a disconnected dot appears for each data point. The overall pattern of these scattered dots often exposes a pattern or a trend.

A simple scatterplot

The following steps show you how to construct a simple scatterplot:

1. **Choose File⇨Open⇨Data and open the Employee data.sav file.**

The file is in the SPSS installation directory.

2. **Choose Graphs⇨Chart Builder.**

3. **In the Choose From list, select Scatter/Dot.**

4. **Select the simplest scatterplot diagram (the one in the upper-left corner of the examples), and drag it to the panel at the top.**

5. **In the Variables list, select Beginning Salary and drag it to the rectangle labeled X-Axis in the diagram.**

6. **In the Variables list, select Current Salary and drag it to the rectangle labeled Y-Axis in the diagram.**

7. **Click the OK button.**

   The chart in Figure 10-4 appears.

Each dot on the scatterplot in Figure 10-4 represents both the starting salary and the current salary of one employee. The most obvious fact you can derive from this is that the current salary depends largely on the starting salary. In the pattern of the dots, it's easy to see a normal line from the lower left to the upper right. Any dot on that imaginary line represents the salary of an employee who received a normal raise. The dots above the line are the employees who got above-average raises, and those below the line are those with below-average raises. This plot has the shortcoming that the length of service is not considered.

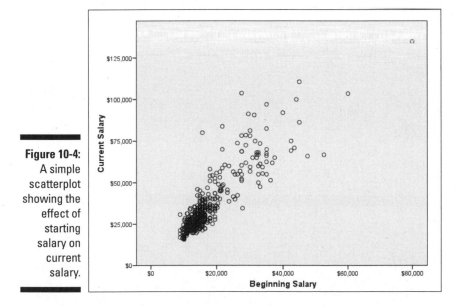

Figure 10-4:
A simple scatterplot showing the effect of starting salary on current salary.

Scatterplot showing multiple variables

You can display more than one variable along the same axis. The following example constructs a scatterplot showing the beginning salary and the current salary according to the number of months of experience the person had before taking the job:

1. **Choose File⇨Open⇨Data and open the `Employee data.sav` file, which is in the SPSS installation directory.**

2. **Choose Graphs⇨Chart Builder.**

3. **In the Choose From list, select Scatter/Dot.**

4. **Select the scatterplot diagram in the top row, center, and drag it to the panel at the top.**

5. **In the Variables list:**

    a. **Select Beginning Salary and drag it to the Y-Axis rectangle.**

    b. **Select Current Salary and drag it to the same location as you dropped the Beginning Salary.**

    Be careful to drop it on the square with the plus sign. The plus sign appears as you drag the droppable item over the rectangle.

    c. **Select Previous Experience and drag it to the X-Axis rectangle.**

6. **Click the OK button.**

    The chart shown in Figure 10-5 appears, with two different colored dots and a legend at the upper right.

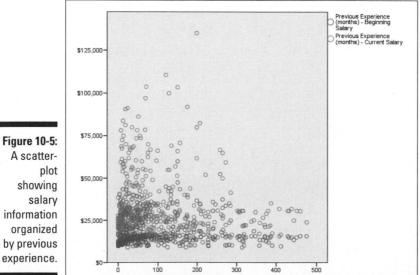

Figure 10-5:
A scatter-plot showing salary information organized by previous experience.

Three-dimensional scatterplot

Three-dimensional scatterplots can be dramatic in appearance, but clarity is not their strongest point. Because the scatterplot is drawn on a two-dimensional surface, you might find it difficult to envision where each point is supposed to appear in space. However, if your data distributes appropriately on the display, the chart may demonstrate the concept you're trying to get across.

The following example uses the same data as in the preceding example but displays it in a different way, as a three-dimensional plot:

1. **Choose File⇨Open⇨Data and open the `Employee data.sav` file.**

   The file is in the SPSS installation directory.

2. **Choose Graphs⇨Chart Builder.**

3. **In the Choose From list, select Scatter/Dot.**

4. **Select the scatterplot diagram in the top row on the right and drag it to the panel at the top.**

5. **In the Variables list:**

   a. Select Beginning Salary and drag it to the X-Axis rectangle.

   b. Select Current Salary and drag it to the Y-Axis rectangle.

   c. Select Previous Experience and drag it to the Z-Axis rectangle.

6. **Click the OK button.**

   The graph shown in Figure 10-6 appears.

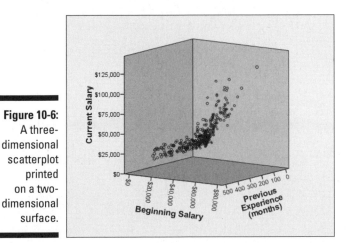

Figure 10-6: A three-dimensional scatterplot printed on a two-dimensional surface.

Dot plot

No plot is simpler to produce than the *dot plot*. It has only one dimension. Although SPSS groups it among the scatterplots, there's nothing scattered about it. It actually presents data more like a bar chart — and it reminds me of that old joke about stacking BBs.

It's easy to create a dot plot. You select the dot plot as the type of graph you want and then select one variable. SPSS does the rest. The following steps guide you through the process of creating a simple dot plot:

1. **Choose File⇨Open⇨Data and open the `Employee data.sav` file.**

   The file is in the SPSS installation directory.

2. **Choose Graphs⇨Chart Builder.**

3. **In the Choose From list, select Scatter/Dot.**

4. **Select the rightmost image in the second row (the one that's three vertical stacks of circular dots) and drag it to the panel at the top.**

   This is the simple dot plot.

5. **In the Variables list, select Date of Birth and drag it to the X-Axis rectangle.**

6. **Click the OK button.**

   The chart shown in Figure 10-7 appears.

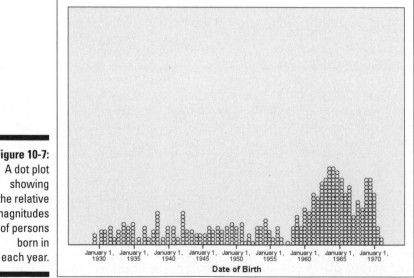

Figure 10-7:
A dot plot showing the relative magnitudes of persons born in each year.

Scatterplot matrix

A *scatterplot matrix* is a group of scatterplots combined into a single graphic. You choose a number of scale variables and include them as a member of your matrix, and SPSS creates a scatterplot for each possible pair of variables. You can make the matrix as large as you like — its size is controlled by the number of variables you include.

The following steps walk you through the creation of a matrix:

1. **Choose File⇨Open⇨Data and open the `Cars.sav` file.**

   The file is in the SPSS installation directory.

2. **Choose Graphs⇨Chart Builder.**

3. **In the Choose From list, select Scatter/Dot.**

4. **Select the windowpane-looking image in the lower-left corner and drag it to the panel at the top.**

5. **In the Variables list, drag Beginning Salary to the Scattermatrix rectangle in the panel at the top.**

   The selected name replaces the label in the rectangle.

6. **In similar fashion, drag the variable names Current Salary, Months since Hire, and Previous Experience (Months) to the rectangle inside the panel at the top of the window.**

   The labels may or may not change with each variable you add (depending on the length and amount of space available), but they will all be added to the list at the bottom of the Element Properties dialog box.

7. **Click the OK button.**

   The chart in Figure 10-8 appears. As you can see, each variable is plotted against each of the others.

The matrix of scatterplots in Figure 10-8 has each variable plotted against each of the others. Notice that the scatterplots along the diagonal from the upper left to the lower right are blank — that's because it's useless to plot a variable against itself. Also, notice the symmetry. All the plots in the lower-left half have a rotated and mirrored image in the upper-right half.

Drop-line chart

A *drop-line chart* presents a special kind of summary with points and vertical lines. The points are grouped horizontally at each categorical value with a line drawn vertically through them. This arrangement can be visually helpful when comparing the values that appear within each category.

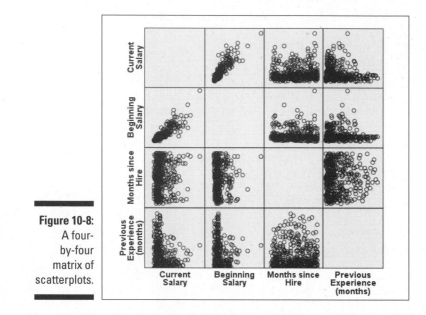

Figure 10-8:
A four-
by-four
matrix of
scatterplots.

The following steps take you through the basic actions necessary for producing a drop-line graph:

1. **Choose File⇨Open⇨Data and open the `Cars.sav` file.**

   The file is in the SPSS installation directory.

2. **Choose Graphs⇨Chart Builder.**

3. **In the Choose From list, select Scatter/Dot.**

4. **Select the image with the vertical line, in the center of the bottom row, and drag it to the panel at the top.**

5. **Select the last of the possible chart options — the diagram showing vertical lines joining open dots — and drag it to the panel at the top.**

6. **In the Variables list:**

   a. **Select Number of Cylinders and drag it to the rectangle in the upper-right corner with the Set Color label.**

   b. **Select Model Year and drag it to the X-Axis rectangle.**

   c. **Select Horsepower and drag it to the Y-Axis rectangle.**

      Note that X-Axis and Set Color both contain categorical variable names, and the Y-Axis contains a scale variable. This is the only combination of variable types that will work.

7. **Click the OK button.**

   The graph in Figure 10-9 appears.

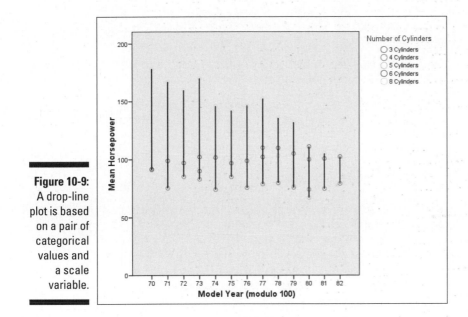

Figure 10-9:
A drop-line
plot is based
on a pair of
categorical
values and
a scale
variable.

Bar Graphs

A *bar graph* is a comparison of relative magnitudes. Simple bar graphs and simple line graphs are the most common ways of charting statistics. It would make an interesting statistical study to determine which is the more common. The results could be displayed as either a bar graph or a line graph, whichever is more popular.

Simple bar graph

A fundamental bar graph is simple enough that the decisions you need to make when preparing one are almost intuitive. The following steps can be used to generate a simple bar graph:

1. **Select File⇨Open⇨Data and open the** `Employee data.sav` **file.**

   The file is in the SPSS installation directory.

2. **Choose Graphs⇨Chart Builder.**

3. **In the Choose From list, select Bar.**

4. **Select the Simple Bar image — the one in the upper-left corner — and drag it to the panel at the top of the window.**

5. **In the Variables list, select Education Level and drag it to the X-Axis rectangle.**

6. **In the Variables list, select Current Salary and drag it to the Count rectangle.**

   The label changes from Y-Axis to Count to indicate the type of variable that should now be applied to that axis.

7. **Click the OK button.**

   The bar graph in Figure 10-10 appears.

Clustered bar chart

A *clustered bar chart* can show the relationships among a cluster of items by displaying more than one value and presenting a summary of categorical values. Clustering combines several bar charts into a single display. The following steps take you through the process of constructing a clustered bar chart:

1. **Choose File⇨Open⇨Data and open the `Cars.sav` file.**

   The file is in the SPSS installation directory.

2. **Choose Graphs⇨Chart Builder.**

3. **In the Choose From list, select Bar.**

4. **Select the Clustered Bar image — the one in the center of the top row — and drag it to the panel at the top of the window.**

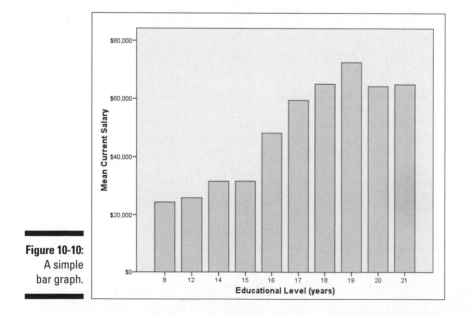

Figure 10-10:
A simple
bar graph.

5. **In the Variables list:**

    a. **Select Model Year and drag it to the X-Axis rectangle.**

    b. **Select Horsepower and drag it to the Count rectangle.**

    The rectangle was originally labeled Y-Axis. The label changed to help you understand the type of variable that needs to be placed there.

    c. **Select Number of Cylinders and drag it to the rectangle in the upper-right corner, the one now labeled Cluster.**

6. **Click the OK button.**

    The graph in Figure 10-11 appears.

Stacked bar chart

A *stacked bar chart* is similar to the clustered bar chart in that it displays multiple values of a variable for each value of a categorical variable. The following chart displays the same data as the preceding example, but emphasizes different aspects of the data.

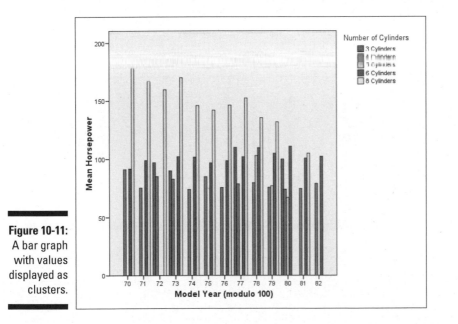

Figure 10-11:
A bar graph with values displayed as clusters.

The following steps can be followed to create a stacked bar chart:

1. **Choose File⇨Open⇨Data and open the `Cars.sav` file.**

   The file is in the SPSS installation directory.

2. **Choose Graphs⇨Chart Builder.**

3. **In the Choose From list, select Bar.**

4. **Select the Stacked Bar image — the one on the right side of the top row — and drag it to the panel at the top of the window.**

5. **In the Variables list:**

   a. **Select Model Year and drag it to the X-Axis rectangle.**

   b. **Select Horsepower and drag it to the Count rectangle.**

      The rectangle was originally labeled Y-Axis. The label changed to help you understand the type of variable that needs to be placed there.

   c. **Select Number of Cylinders and drag it to the rectangle in the upper-right corner, the one now labeled Stacks.**

6. **Click the OK button.**

   The graph in Figure 10-12 appears.

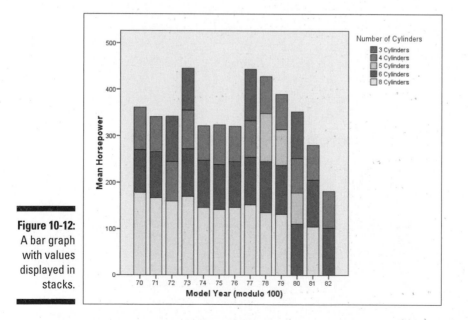

Figure 10-12:
A bar graph with values displayed in stacks.

Three-dimensional bar chart

A simple *three-dimensional bar chart* is the same as a two-dimensional bar chart, except a third variable is added to specify the values along the new dimension. As with most three-dimensional displays, it has the advantage of displaying three relative values at once, and it has the disadvantage of making it difficult to determine which is the greater of two values if the two values are close.

The following steps construct a three-dimensional bar chart:

1. **Choose File⇨Open⇨Data and open the `Cars.sav` file.**

   The file is in the SPSS installation directory.

2. **Choose Graphs⇨Chart Builder.**

3. **In the Choose From list, select Bar.**

4. **Select the Simple Three Dimensional Bar image — the one on the left end of the second row — and drag it to the panel at the top of the window.**

5. **In the Variable list:**

   a. **Select Model Year and drag it to the Y-Axis rectangle.**

   b. **Select Number of Cylinders and drag it to the X-Axis rectangle.**

   c. **Select Country of Origin and drag it to the Z-Axis rectangle.**

6. **Click the OK button.**

   The graph in Figure 10-13 appears.

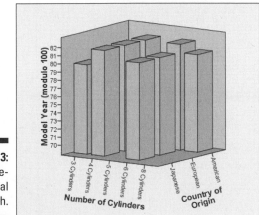

Figure 10-13:
A three-dimensional bar graph.

Error bars

Some errors come from flat-out mistakes. That's not the kind of error I talk about here. Statistical sampling can help you arrive at a conclusion, but that conclusion has a margin of error. This margin can be calculated and quantified according to the size of the sample and the distribution of the data. For example, suppose that you want to know how typical the result is when you calculate the mean of all the values for some variable — for any one case the value could be as much as the largest value or as small as the smallest. The maximum and minimum are the extremes of the possible error. You can choose values and mark the points that contain, say, 90 percent of all values. Marking these points on graphs creates *error bars*.

You can add error bars to the display of most types of graphs. For example, you could add error bars to the simple bar graph presented earlier in this chapter (refer to Figure 10-10) by making selections in the Element Properties dialog box. If you've worked through any of the examples, you'll know Element Properties as that pesky window that pops up every time you construct a chart.

For an example of adding error bars to a bar chart, follow the same procedure described previously in the "Simple bar graph" section, but just before the final step (clicking the OK button to produce the chart), do the following:

1. **If the Element Properties window is not displayed, click the Element Properties button.**

2. **In the Element Properties window, make sure that a check mark appears in the Display Error Bars option.**

3. **Select Confidence Level Intervals and set its value to 95%.**

4. **Click the Apply button.**

5. **Click the OK button.**

   The chart in Figure 10-14 is displayed.

You can display the range of errors without displaying the full bars. To do this with the same data as before, perform the following steps:

1. **Choose File⇨Open⇨Data and open the `Employee data.sav` file.**

   The file is in the SPSS installation directory.

2. **Choose Graphs⇨Chart Builder.**

3. **In the Choose From list, select Bar.**

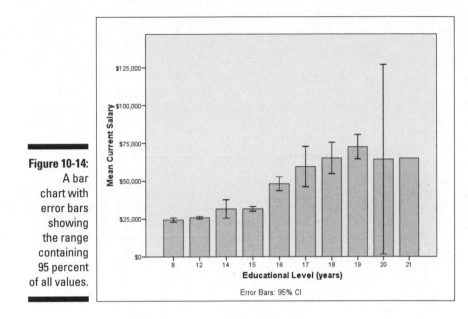

Figure 10-14:
A bar
chart with
error bars
showing
the range
containing
95 percent
of all values.

4. **Select the Simple Error Bar image — the one in the bottom row on the left — and drag it to the panel at the top of the window.**

5. **In the Variables list, select Education Level and drag it to the X-Axis rectangle.**

6. **In the Variables list, select Current Salary and drag it to the Mean rectangle.**

   The label changes from Y-Axis to Mean to indicate the type of data that will be displayed on that axis.

7. **In the Element Properties window, make sure that the Display Error Bars option is checked, the Confidence Intervals is selected, and the Level is set to 95%.**

8. **Click the OK button.**

   The bar graph in Figure 10-15 appears.

This example displays the result of one way of making error calculations. In this example, the magnitude of the error is based on 95 percent of all values being within the upper and lower error bounds. You can base the error also on the bell curve and mark the upper and lower errors at some multiple of the standard error or standard deviation.

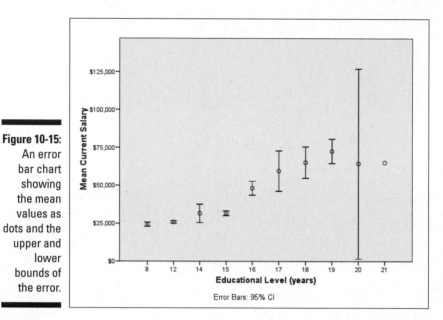

Figure 10-15:
An error
bar chart
showing
the mean
values as
dots and the
upper and
lower
bounds of
the error.

Anytime you make a change to a setting or a value in the Element Properties dialog box, you must click the Apply button to have the change reflected in your chart.

Chapter 11

More Types of Graphs

SPSS has a number of ways to present data graphically. This chapter, and the one before it, provide an overview of many of the charts available. Some are more appropriate than others for different kinds of data. Every example in these two chapters is as simple as possible to present you with a general idea of the types of charts you can choose from. Remember that this is only a representative selection. You start by choosing a basic form, and then continue by setting options to display your data in the best way possible. The Element Properties window, which appears automatically, provides you with every possible option that applies to the chart you're building.

When using Chart Builder, it is completely safe to drag and drop any variables you want to see in your graph — if the variable doesn't fit there, the drop will fail. SPSS does you the kindness of figuring out what will and won't work. Also, no matter what you try to do while building a graph, your data will never be hurt.

Histograms

A *histogram* represents the number of items that appear within a range (or within a bin, statistically speaking). You can use a histogram to look at a graphic representation of the frequency distribution of the values of a variable. Histograms are useful for demonstrating the patterns in your data when you want to display information to others rather than discover data patterns for yourself.

Simple histogram

You can use the following steps to create a simple histogram that displays the number of automobiles, in the survey used in the example, having various gas mileage capabilities for each of several years:

1. **Choose File⇨Open⇨Data and open the `Cars.sav` file, which is in the SPSS installation directory.**

2. **Choose Graphs⇨Chart Builder.**

   The Chart Builder dialog box appears.

3. **In the Choose From list, select Histogram.**

4. **Drag the diagram on the left of the top row to the panel at the top of the window.**

5. **In the Variables list:**

   a. **Select the Model Year variable and drag it to the Y-Axis rectangle in the panel.**

   b. **Select Miles Per Gallon and drag it to the Count rectangle in the panel.**

6. **Click the OK button.**

   The histogram shown in Figure 11-1 appears.

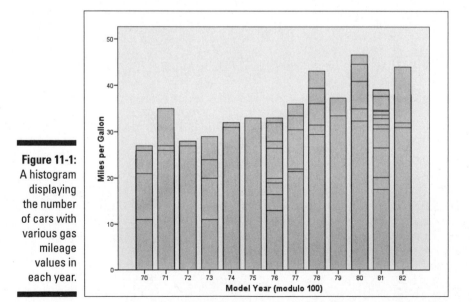

Figure 11-1:
A histogram displaying the number of cars with various gas mileage values in each year.

The graph in Figure 11-1 looks like a bar chart, but it isn't. The height of each bar does not represent the mean or an average — the height is determined by the largest value. The lines drawn across each bar represent the various values of gas mileage in that year. The meaning of a graph of this sort is not intuitive and probably should be accompanied by a note explaining what it means.

Stacked histogram

You can create a histogram that is more like a bar chart and more intuitive than a simple histogram. In a *stacked histogram,* the overall height of the bars represents the mean of the values in each category, and different categories of a third variable are indicated by displaying portions of the bars in different colors.

The following steps produce a stacked histogram displaying the same values as the preceding simple histogram, plus the number of cylinders:

1. **Choose File⇨Open⇨Data and open the `Cars.sav` file.**

2. **Choose Graphs⇨Chart Builder.**

3. **In the Choose From list, select Histogram.**

4. **Drag the diagram in the center of the top row to the panel at the top of the window.**

5. **In the Variables list:**

   a. **Select the Model Year variable and drag it to the Y-Axis rectangle.**

   b. **Select Miles Per Gallon and drag it to the Count rectangle.**

   c. **Select Number of Cylinders and drag it to the Stack rectangle, in the upper-right.**

6. **Click the OK button.**

   The histogram shown in Figure 11-2 appears.

In this type of histogram, the scale on the left became the mean of the value, which means the overall height of each bar is, like a bar chart, the mean of the miles per gallon in each model year. Each bar is comprised of a stacking of rectangles representing the portion of the total that was made up from cars with a certain number of cylinders. The overall area of each bar and the area of each rectangle making up the bar represent the mean.

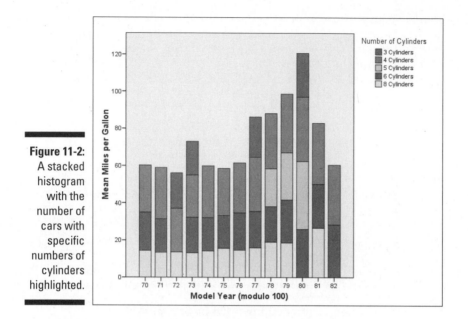

Figure 11-2:
A stacked histogram with the number of cars with specific numbers of cylinders highlighted.

Frequency polygon

The *frequency polygon* is a histogram that looks like a line chart. It also looks like an area graph, which is described in the next section. A frequency polygon is as easy to construct as a simple histogram. The following steps guide you through a procedure that produces a frequency polygon histogram:

1. **Choose File⇨Open⇨Data and open the `Cars.sav` file.**

2. **Choose Graphs⇨Chart Builder.**

3. **In the Choose From list, select Histogram.**

4. **Drag the diagram on the right end of the top row to the panel at the top of the window.**

5. **In the Variables list:**

    a. **Select the Model Year variable and drag it to the X-Axis rectangle in the panel.**

    a. **Select Miles Per Gallon and drag it to the Y-axis rectangle.**

6. **Click the OK button.**

    The histogram shown in Figure 11-3 appears.

The frequency polygon is the simplest histogram of them all. It is a representation of the mean of the value on the Y-axis, so it does not really give you an idea of the relative number of items in each category.

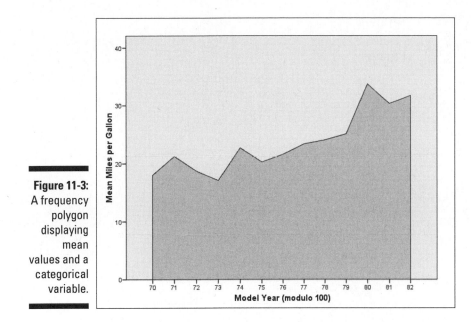

Population pyramid

A *population pyramid* provides an immediate comparison of the number of items that fall into categories. It is called a pyramid because it often takes that shape — wide at the bottom and tapering to a point at the top. The following steps can be followed to build an example pyramid histogram chart:

1. **Choose File⇨Open⇨Data and open the Employee data.sav file, which is in the SPSS installation directory.**

2. **Use the tab to switch to Variable View.**

3. **Select the Type column of the bdate variable.**

4. **Click the button that appears near the variable type name, which is Date.**

5. **In the list of date formats, choose mmm yyy and then click the OK button.**

   This is a matter of personal preference. The chart is produced no matter which format is used to display the dates, but I think this format looks better than most of the others.

6. **Choose Graphs⇨Chart Builder.**

7. **In the Choose From list, select Histogram.**

8. **Drag the blue and green chart in the second row to the panel at the top of the window.**

9. **In the Variables list:**

   a. **Select the Gender variable and drag it to the Split Variable rectangle.**

      This is a categorical variable with two possible values, so one category will be placed on each side of the center line.

   b. **Select Date of Birth and drag it to the Distribution Variable rectangle.**

10. **Click the OK button.**

    The chart shown in Figure 11-4 appears.

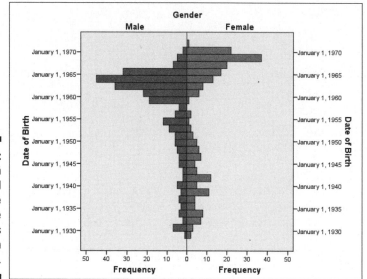

Figure 11-4: A population pyramid shows the occurrence of values within categories.

You can create pyramid histograms based on categorical variables with three, four, or more values. The plot produced will consist of as many pairs (and one single-sided pyramid, if necessary) as needed to display bars showing the relative number of occurrences of different values in the categories.

Area Graphs

An *area graph* is really a line graph, or a collection of line graphs, with areas below the lines filled in to represent the mean of one or more values at the various points of the other axis.

Simple area graph

A *simple area graph* displays the area below a single line. The following steps will produce a simple area graph:

1. **Choose File⇨Open⇨Data and open the `Employee data.sav` file, which is in the SPSS installation directory.**

2. **Choose Graphs⇨Chart Builder.**

3. **In the Choose From list, select Area.**

4. **Drag the diagram on the left — the one with a single line — to the panel at the top of the window.**

5. **In the Variables list:**

    a. **Select the Educational Level variable and drag it to the X-Axis rectangle.**

    b. **Select Beginning Salary and drag it to the Count rectangle.**

       This is the rectangle that was labeled Y-Axis until the X-Axis became defined.

6. **Click the OK button.**

    The area chart shown in Figure 11-5 appears.

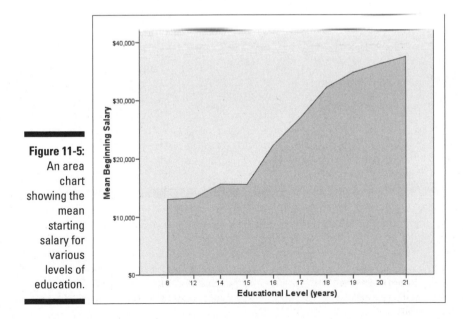

Figure 11-5:
An area chart showing the mean starting salary for various levels of education.

Stacked area chart

A *stacked area chart* is a chart with more than one variable being calculated along the X-axis. The values are stacked in such a way that the ups and downs of the lower value in the chart have an effect on the upper values in the chart. That is, the charting is not a group of independent lines but is, instead, a representation of a cumulative total with the value added by each variable displayed.

Follow these steps to produce a stacked area chart:

1. **Choose File➪Open➪Data and open the `Employee data.sav` file.**

2. **Choose Graphs➪Chart Builder.**

3. **In the Choose From list, select Area.**

4. **Drag the diagram on the right — the one with multiple lines — to the panel at the top of the window.**

5. **In the Variables list:**

   a. **Select the Educational Level variable and drag it to the X-Axis rectangle.**

   b. **Select Current Salary and drag it to the Count rectangle.**

   c. **Select Beginning Salary and drag it to the Current Salary rectangle.**

   Make certain you drag it to the plus sign and not simply to the rectangle in general. (The plus sign appears at the top of the rectangle when you drag the new variable name across it.)

6. **Click the OK button.**

   The area chart shown in Figure 11-6 appears.

You can drag and stack a number of variables. They all appear in the legend at the upper right, and each variable makes one layer of the stack.

It is important that the variables you select for stacking have similar ranges of values so that the scale on the left side will make sense for all of them. If, for example, one variable ranges into the thousands and the other doesn't get over a hundred, the smaller one will compress and come out in the final graph as a line.

The variables you select to be stacked must be selected in the order you want to stack them. That is, the first one you select will be on top. The second one you select will be next to the top, and so on.

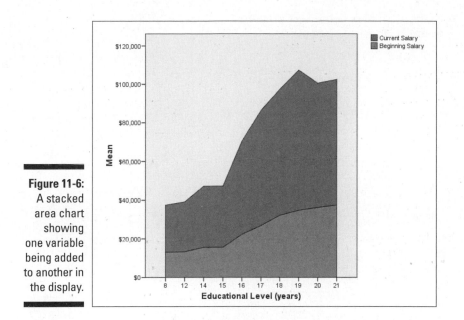

Figure 11-6:
A stacked area chart showing one variable being added to another in the display.

The two types of area charts, simple and stacked, act the same. You can select the stacked chart and produce a single-area chart, or you can start with the simple area chart and stack your variables.

Pie Charts

Pie charts are the easiest kind to spot — they are the only charts in circles. The purpose of a *pie chart* is simply to show how something (the "whole") is divided into pieces. You can divide something into two pieces, ten pieces, or any other number. Each slice in the pie chart represents its percentage of the whole. For example, if a slice takes up 40 percent of the total pie, that slice represents 40 percent of the total number. A pie chart is also called a polar chart.

In the following steps, you construct a simple pie chart:

1. **Choose File⇨Open⇨Data and open the `Employee data.sav` file, which is in the SPSS installation directory.**

2. **Choose Graphs⇨Chart Builder.**

3. **In the Choose From list, select Pie/Polar.**

4. **Drag the pie diagram to the panel at the top of the window.**

5. **In the Variables list, drag Educational Level to the Slice By rectangle at the bottom of the panel.**

6. **Click the OK button.**

   The pie chart shown in Figure 11-7 appears.

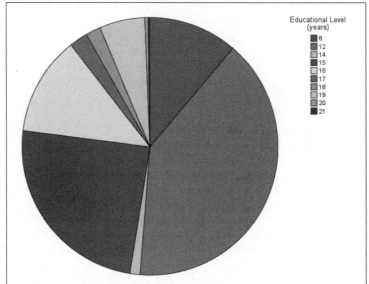

Figure 11-7: A pie chart displaying the number of employees at each education level.

Boxplots

A *boxplot* uses graphic elements to display five statistics at one time within each categorical value. The statistics are the minimum value, first quartile, median value, third quartile, and maximum value. A boxplot is particularly good for helping you spot values lying well outside the range of normal values.

Simple boxplot

A one-dimensional boxplot displays the range of values for all cases for one categorical variable. The following steps guide you through the creation of a one-dimensional boxplot:

1. **Choose File⇨Open⇨Data and open the Employee data.sav file, which is in the SPSS installation directory.**

2. **Choose Graphs⇨Chart Builder.**

3. **In the Choose From list, select Boxplot.**

4. **Drag the diagram on the left to the panel at the top of the window.**

5. **In the Variables list:**

   a. **Select the Educational Level variable and drag it to the X-Axis rectangle.**

   b. **Select the Current Salary variable and drag it to the Y-Axis rectangle.**

6. **Click the OK button.**

   The boxplot shown in Figure 11-8 appears.

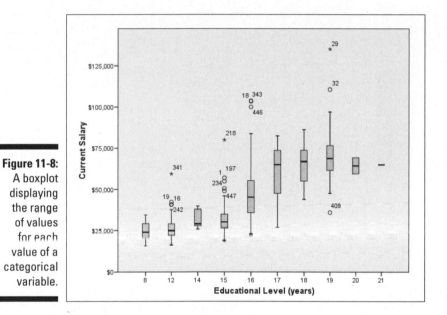

Figure 11-8:
A boxplot displaying the range of values for each value of a categorical variable.

In Figure 11-8, each vertical column of graphics represents all the values for a category. The values marked with either circles or stars are the ones beyond the extents of the first and third quartiles. The ones marked by stars are the extremes. You can look at a boxplot of this type to find data out of whack.

Clustered boxplot

A *clustered boxplot* displays the values of three variables at once. Use the following steps to construct a clustered boxplot:

1. **Choose File⇨Open⇨Data and open the Employee data.sav file.**

2. **Choose Graphs⇨Chart Builder.**

3. **In the Choose From list, select Boxplot.**

4. **Drag the diagram from the center of the row to the panel at the top of the window.**

5. **In the Variables list:**

   a. **Drag the Minority Classification variable to the X-Axis rectangle.**

   b. **Drag the Current Salary variable to the Y-Axis rectangle.**

   c. **Drag the Educational Level variable to the Cluster rectangle.**

6. **Click the OK button.**

   The boxplot shown in Figure 11-9 appears.

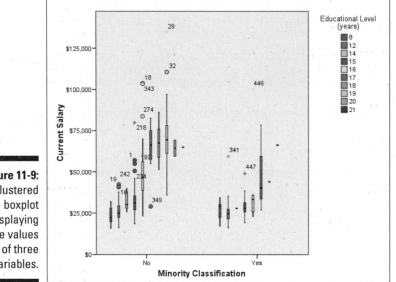

Figure 11-9: A clustered boxplot displaying the values of three variables.

A boxplot displays a lot of information. With three variables being displayed, it can get very busy. It is actually easier to read on the screen than it is here on this page in shades of gray. The legend in the upper-right corner assigns colors to the categorical values, and those colors appear in the boxes to show you which is which. You are also shown the ID numbers of cases with extreme values.

One-dimensional boxplot

A *one-dimensional boxplot* displays one variable in such a way that you can easily see the range of values and spot out-of-range values. The following steps construct an example of a one-dimensional boxplot:

1. **Choose File⇨Open⇨Data and open the `Employee data.sav` file.**

2. **Choose Graphs⇨Chart Builder.**

3. **In the Choose From list, select Boxplot.**

4. **Drag the diagram on the right end of the row to the panel at the top of the window.**

5. **Click the Groups/Point ID tab and select the Point To ID Label option.**

   A rectangle labeled Point ID Variable appears in the upper-right corner of the panel at the top.

6. **In the Variables list:**

   a. **Drag the Employee Code variable to the new rectangle in the upper right of the panel.**

   b. **Drag the Current Salary variable to the X-Axis rectangle, on the left side of the panel.**

7. **Click the OK button.**

   The boxplot shown in Figure 11-10 appears.

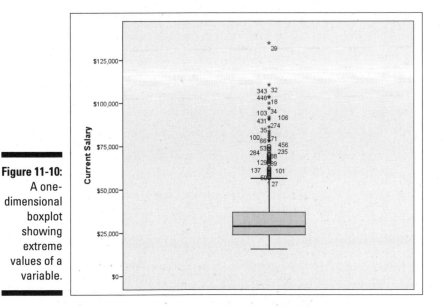

Figure 11-10: A one-dimensional boxplot showing extreme values of a variable.

The boxplot in Figure 11-10 graphically displays values out of the normal range. Each value is tagged with the ID number of its case. The number displayed as the ID is the variable previously chosen as the point ID. If no point ID variable had been chosen, the annotation shows the normal SPSS case numbers.

High-Low Graphs

A *high-low chart* displays the range of values between specified high and low values. Its purpose is to compare two or three variables.

High-low close

The *high-low close graph* shows how a variable appears when plotted between a high value and a low value. That is, it displays the relationships among three sets of values. This example and the one that follows display the same information, but with a different layout of the graphics.

Follow these steps:

1. **Choose File⇨Open⇨Data and open the file named `Home sales [by neighborhood].sav`, which is in the SPSS installation directory.**

2. **Choose Graphs⇨Chart Builder.**

3. **In the Choose From list, select High-Low.**

4. **Drag the diagram on the left of the top row to the panel at the top of the window.**

5. **In the Variables list:**

   a. **Drag the Employee Code variable to the new rectangle in the upper right of the panel.**

   b. **Drag the Neighborhood variable to the X-Axis rectangle.**

   c. **Drag the Select Sale Price variable to the Close Variable rectangle.**

   d. **Drag the Select Appraised Land Value variable to the Low Variable rectangle.**

   e. **Drag the Select Total Appraised Value variable to the High Variable rectangle.**

6. **Click the OK button.**

   The high-low graph shown in Figure 11-11 appears.

Simple range bar

The *simple range bar* graph shows how a variable appears when plotted between high and low values. That is, it displays the relationships among three sets of values. This example and the one before it display the same information, but with a different layout of the graphics.

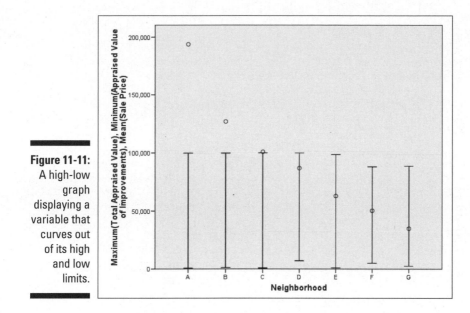

Figure 11-11:
A high-low graph displaying a variable that curves out of its high and low limits.

Do the following to build a simple range bar graph:

1. **Choose File⇨Open⇨Data and open the `Home sales [by neighbor-hood].sav` file, which is in the SPSS installation directory.**

2. **Choose Graphs⇨Chart Builder.**

3. **In the Choose From list, select High-Low.**

4. **Drag the diagram in the center of the top row to the panel at the top of the window.**

5. **In the Variables list:**

   a. **Drag the Employee Code variable to the new rectangle in the upper right of the panel.**

   b. **Drag the Neighborhood variable to the X-Axis rectangle.**

   c. **Drag the Select Sale Price variable to the Close Variable rectangle.**

   d. **Drag the Select Appraised Land Value variable to the Low Variable rectangle.**

   e. **Drag the Select Total Appraised Value variable to the High Variable rectangle.**

6. **Click the OK button.**

   The high-low graph shown in Figure 11-12 appears.

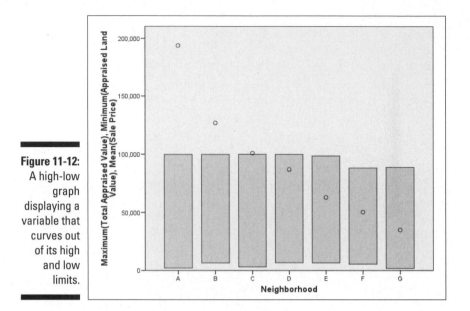

Figure 11-12:
A high-low
graph
displaying a
variable that
curves out
of its high
and low
limits.

Differenced area

A *differenced area graph* provides a pair of line graphs of variables with their differences emphasized by filling the area between the two with a solid color. The two graphs are plotted against the points of a categorical variable. The following steps produce a differenced area graph:

1. **Choose File⇨Open⇨Data and open the Home sales [by neighborhood].sav file, which is in the SPSS installation directory.**

2. **Choose Graphs⇨Chart Builder.**

3. **In the Choose From list, select High-Low.**

4. **Drag the diagram from the second row to the panel at the top of the window.**

5. **In the Variables list:**

   a. **Drag the Neighborhood variable to the X-Axis rectangle.**

   b. **Drag the Select Sale Price variable to either of the Y-Axis rectangles.**

   c. **Drag the Select Appraised Value of Improvements variable to the other Y-Axis rectangle.**

6. **Click the OK button.**

   The differenced area chart shown in Figure 11-13 appears.

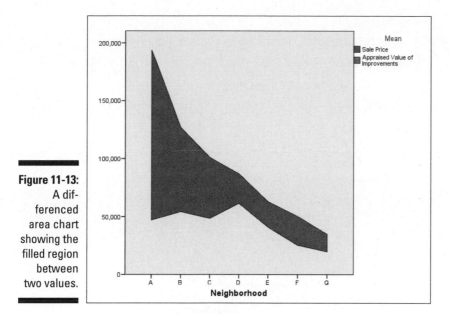

Figure 11-13:
A differenced area chart showing the filled region between two values.

Dual-Axis Graphs

Many of the other graphic forms allow you to plot two or more variables on the same chart, but they must always be plotted against the same scale. In the dual-axis graph, two variables are plotted and two different scales are used to plot them. As a result, the values don't require the same ranges, as they do in the other plots, and the curves and trends of the two variables can be easily compared.

Dual Y-axes with categorical X-axis

Two variables with different ranges that vary across the same set of categories can be plotted together, as shown in the following example:

1. **Choose File⇨Open⇨Data and open the Cars.sav file, which is in the SPSS installation directory.**

2. **Choose Graphs⇨Chart Builder.**

3. **In the Choose From list, select Dual Axes.**

4. **Drag the diagram on the left to the panel at the top of the window.**

5. **In the Variables list:**

   a. **Drag the Horsepower variable to the Y-Axis rectangle.**

   b. **Drag the Miles Per Gallon variable to the Y-Axis rectangle.**

   c. **Drag the Number of Cylinders variable to the X-Axis rectangle.**

6. **Click the OK button.**

   The dual-axis graph shown in Figure 11-14 appears.

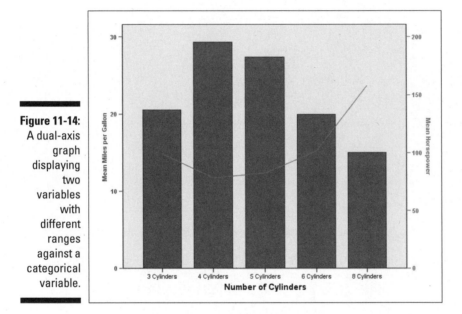

Figure 11-14:
A dual-axis graph displaying two variables with different ranges against a categorical variable.

Dual Y-axes with scale X-axis

Two variables with different ranges that vary according to the changes in a third scale value can be plotted together, as shown in the following example:

1. **Choose File⇨Open⇨Data and open the `Cars.sav` file, which is in the SPSS installation directory.**

2. **Choose Graphs⇨Chart Builder.**

3. **In the Choose From list, select Dual Axes.**

4. **Drag the diagram on the right to the panel at the top of the window.**

5. **In the Variables list:**

    a. **Drag the Miles Per Gallon variable to the Y-Axis rectangle.**

    b. **Drag the Engine Displacement variable to the Y-Axis rectangle.**

    c. **Drag the Select Time to Accelerate 0 to 60 variable to the X-Axis rectangle.**

6. **Click the OK button.**

    The dual-axis chart shown in Figure 11-15 appears.

Figure 11-15:
A dual-axis graph displaying two variables with different ranges against a scale variable.

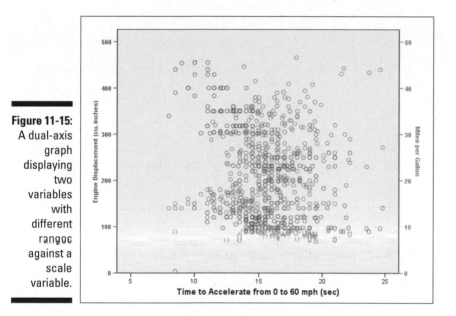

The graph displayed in Figure 11-15 is a combination of two dot-plot formats, with the dots in different colors. Even on a color display, the two sets of values — each set plotted on a different Y-axis scale — can be confusing. With this type of plot, you must take care that your data makes sense being displayed this way.

Chapter 12

Maps

In This Chapter

▶ Displaying your data geographically

▶ Making a large volume of data visible at a glance

▶ Choosing map colors and symbols

---

*T*hematic mapping is displaying statistical information on a geographical map. Maps can be color-coded and shaded and can contain special symbols and charts. Most types of maps are made to display relative magnitudes, but some can display exact values.

This chapter shows you how to create different kinds of maps in SPSS. Each kind of map presents data in its own way. After you get an idea of the options, you'll have a good idea about which will show your data at its best.

Relating Geography and Data

A statistical map, or what the SPSS documentation calls a *thematic map,* is a geographical display that displays numeric values assigned to each named region of the map. To match the data with the map, the named regions on the map must be related to names of variables in which the data is stored — that way, the data from each case (row) can be associated with specific areas of the map. After the associations are established, SPSS can graphically display the numeric values for each named region.

You need the following to render a thematic map:

✔ The map in a file format that SPSS can use

✔ Data that contains geographical location information

✔ A definition of a method by which the statistical data will be presented on the map

A map file in the correct format is called a *geoset*. SPSS supplies a number of map files, and you can get more of them here:

```
http://www.spss.com
```

Also you can use Geoset Manager to create maps of your own and edit the maps you already have. If you want, you can customize, add, or delete the layers of a map, but you'll need to be able to execute Geoset Manager, which is named `geosetmanager40.exe` and is in the SPSS installation directory. It's an add-on to SPSS, so you may not have it. However, you have the basic map files whether or not you have Geoset Manager.

You'll need to use your data to specify a geographic location for each case. If you want, you can use *X/Y binding,* which is a pair of variables containing longitude and latitude. You can also use a *point reference table,* which requires that a variable contain a value that can be looked up in a table supplying the X/Y coordinates. The most common method, and the one used in the examples in this chapter, is to have a string variable contain the geographic names.

An example of a string variable that can be used for mapping is shown in Figure 12-1. The `state` variable contains names of states. It could just as well contain ZIP Codes, the names of cities, or other geographical identities, but they must match the ones in the map file.

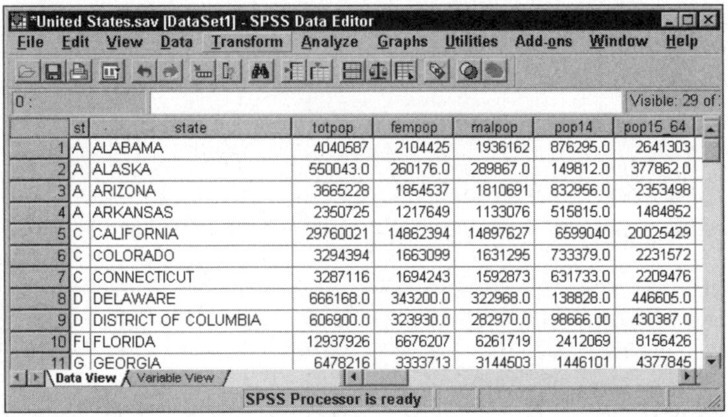

Figure 12-1: A collection of cases with a geographic variable containing place-names.

It's possible to get a geographic mismatch by entering a name in your data that doesn't match one in the map. If you have such a mismatch, SPSS helps you find out where you went wrong. The names that didn't match appear in a list adjacent to the output map. If you double-click anywhere on the map, the map becomes selected and the toolbar shown in Figure 12-2 appears at the top.

Figure 12-2:
The toolbar
that appears
at the top
of a map
when you
double-click.

Figure 12-2:
The toolbar
that appears
at the top
of a map
when you
double-click.

The tool next to the end on the right is the Map Layers tool. Clicking the Map Layers tool displays the Layer Control window shown in Figure 12-3.

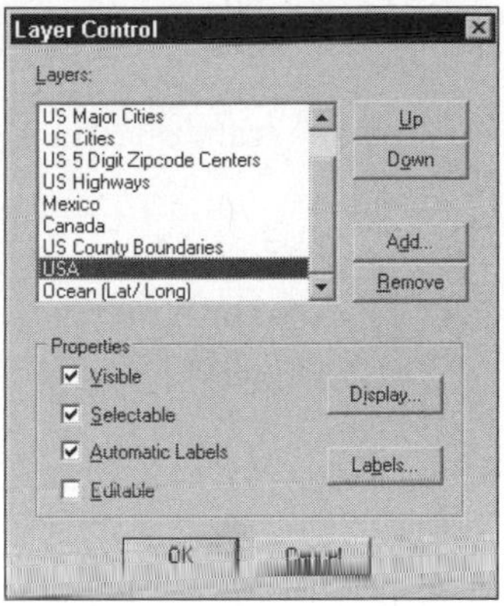

Figure 12-3:
The Layer
Control
window with
settings for
each map
layer.

In the Layer Control window, you can select the layer containing the geographic names your data will have to match. Make certain a check mark appears next to Automatic Labels. Click the OK button, and the labels appear on the map, making it possible for you to see the mistake in your data.

TIP

You may need to zoom in and out to see all the names in a crowded map. To zoom in and out, use the plus and minus sign tools on the map's toolbar (refer to Figure 12-2).

TIP

Using the Layer Control window, you can make various layers visible and invisible to change the appearance of the map. The combinations are almost endless. I suggest experimenting with the options in the window to discover what it can do.

The third part of presenting data on a map is the form of the graphic that displays it. That's what the rest of this chapter is about. The following sections describe the requirements and procedures for displaying data.

Range of Values

Scale variables can have each of their values placed into an upper and lower range, and the range can be indicated by a color. It's sort of a graphic form of binning that makes it easy to see geographic patterns. The following steps create a map displaying values in ranges:

1. **Choose File⇨Open⇨Data and open the `United States.sav` file.**

   The file is in the `MapData` subdirectory of the SPSS directory.

2. **Choose Graphs⇨Map⇨Range of Values.**

   The Create Range of Values Map dialog box appears, as shown in Figure 12-4.

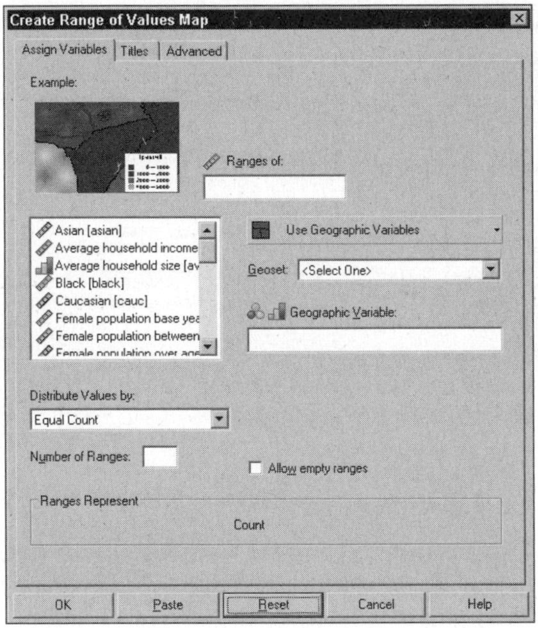

Figure 12-4:
The Create Range of Values Map used to define a map to be drawn.

3. **In the list of variable names on the left, select Total Families, and drag it to the Ranges Of text window on the right.**

   This variable contains the values to be displayed on the map.

4. **Select the State Name variable and drag it to the Geographic Variable text box.**

   This variable contains the location information needed for map placement.

5. **From the Geoset pull-down list, select United States.**

   This is the map to be drawn.

6. **Click the OK button.**

   The map shown in Figure 12-5 appears.

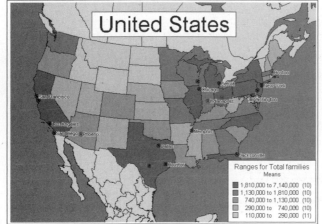

The legend at the lower-right corner of the map tells you which color repre
sents which range of values. The ranges were arrived at automatically by SPSS.
SPSS comes up with what it thinks is a reasonable number of bins (ranges), and
then places each variable into one of the bins. If you want a different number
of bins, enter that number in the Number of Ranges option (refer to Figure 12-4).
Note that the maximum is 5, and that's how many SPSS chose by default in
this example.

You can have SPSS allow empty ranges. If you select a number of ranges and
don't allow empty ranges but a range turns up empty during the rendering of
the map, your selection of the number of ranges is ignored and SPSS goes to
its default.

The setting you choose for Distribute Ranges By specifies how SPSS will construct the range values. The default is Equal Count, where as close as possible to the same number of cases falls into each range (the number within a
range appears in parentheses on the right side of the legend in the drawn
map). Another option is to choose Equal Size, where each range is the same
size. You can select Natural Break and have SPSS look for natural divisions
between groups of values — your data must come in clumps for this option
to be of much use. You can also choose Standard Deviation and have each
range represent the extent of one standard deviation, with the mean value
midway between the two.

At the bottom of the screen, you can select what the ranges represent. The default is to use the mean and distribute things accordingly, but you can select the maximum values, minimum values, variance, mean, median, mode, or a comparison of the numbers of cases.

Click the Titles tab in the dialog box in Figure 12-4 to change the name of the map and the text that appears at the top of the legend.

The Advanced tab presents you with a window for choosing a refining variable for the geography. (For example, you could include a variable that contains the name of a county for each state.) Also, the map is composed of layers, such as county boundaries, and you can turn them off using the Advanced options.

Dot Density

If you need to display graphically which areas have more of something than others, you can represent magnitudes using dot density. The individual dots are almost too small to see, but a group of dots casts a darkness over a region of the map, and that darkness up against the darkness of other areas gives you good notion of relative magnitudes.

The following steps produce a shaded map indicating population density:

1. **Choose File⇨Open⇨Data and open the `United States.sav` file.**

   The file is in the `MapData` subdirectory of the SPSS directory.

2. **Choose Graphs⇨Map⇨Dot Density.**

   The Create Dot Density Map dialog box appears, as shown in Figure 12-6.

3. **In the variable list on the left, select Total Families, and drag it to the Dot Density For text window.**

   This variable contains the values to be displayed on the map.

4. **Select the State Name variable and drag it to the Geographic Variable text box.**

   This variable contains the location information needed for map placement.

5. **In the Geoset pull-down list, select United States.**

   This is the map to be drawn.

6. **Select the Data Value Per Dot radio button and set the value to 1000.**

   The smaller the amount each dot represents, the greater the number of dots placed on the map.

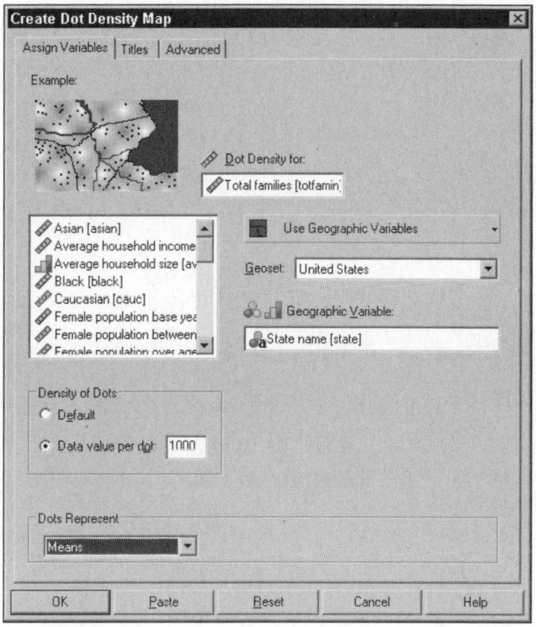

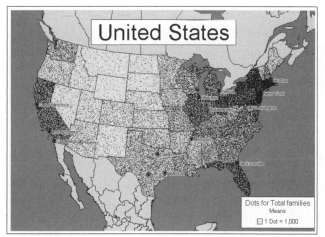

Figure 12-6:
The Create
Dot Density
Map dialog
box used to
define the
map to be
drawn.

7. **Click the OK button.**

   The map shown in Figure 12-7 appears.

Figure 12-7:
A map of
the United
States with
population
represented
by dot
density.

The Dots Represent option in Figure 12-6 tells SPSS how to combine values from different cases. That setting has no effect on this map because there is only one case per state. I suggest that you experiment with this dot density map because you can get dramatic changes in the map's appearance with variations in the number each dot represents.

Double-click the map and a toolbar appears. Select the plus sign and the cursor changes appearance. Place this cursor on an area of interest and click once; the map expands to give you a closer look at the area. Notice that the dots themselves do not expand — they stay the same size, but the distance between them increases because the same number of dots are displayed for each state (even though you may not be able to see them all because some will be off the screen).

You can select the hand icon on the toolbar and use it to move the map around the display. If you move the map in such a way that new parts of it are exposed, you have to wait a few seconds while SPSS draws the newly exposed section. To zoom out, select the minus sign on the toolbar.

Graduated Symbol

You can have symbols on a map represent values, with the size of the symbol indicating the magnitude of the value. A simple example is a map with each state containing a symbol displaying the size of the population relative to the other states. You can construct such a map with the following steps:

1. **Choose File⇨Open⇨Data and open the United States.sav file.**

    The file is in the MapData subdirectory of the SPSS directory.

2. **Choose Graphs⇨Map⇨Graduated Symbol.**

    The Create Graduated Symbol Map shown in Figure 12-8 appears.

3. **In the variable list on the left, select Total Population Base Year, and drag it to the Symbols For text window on the right.**

    This variable determines the symbol sizes on the map.

4. **Select the State Name variable and drag it to the Geographic Variable text box.**

    This variable contains the location information needed for map placement.

5. **In the Geoset pull-down list, select United States.**

    This is the map to be drawn.

Figure 12-8:
The Create
Graduated
Symbol Map
used to
define the
map to be
drawn.

6. **Click the OK button.**

The map shown in Figure 12-9 appears.

Figure 12-9:
A map of
the United
States
with the
population
shown by
the symbol
size.

The default symbol, as you can see in Figure 12-9, is a simple circle. And the default colors make the whole thing look like wads of bubble gum in a puddle of melted pistachio ice cream. But you can change all that with the following steps:

1. **Double-click the map, and the toolbar appears (refer to Figure 12-2).**

2. **Select the third tool from the right — the one named Themes — and a dialog box listing all the themes pops up.**

   With this map, the only theme in the list is the one named Graduated Symbols.

3. **Select the Graduated Symbols name from the list and click the Display button.**

   The Graduated Symbols Theme Options dialog box in Figure 12-10 appears. This is where you make changes to the symbols.

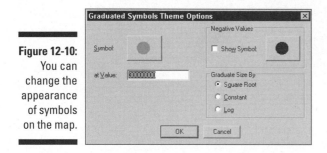

Figure 12-10:
You can change the appearance of symbols on the map.

4. **Click the current symbol, which is in the upper left of the dialog box.**

   A Symbol Styles dialog box appears that exposes the secret of the symbols. The symbols are really characters from the standard fonts, and you can choose any character from any of the fonts listed. You can also select the color of the symbol.

5. **Make your selections from the Symbol Styles dialog box.**

   From the thousands of choices, you can select any symbol from any font, and you can select its color.

Figure 12-11 is the same map with a symbol from the Wingdings font and a different color.

You can tell from the legends on the maps that I lowered the At Value amount in the second map — the size/amount ratio of the symbol — which made the symbols larger. You can also change the relative sizes by changing the Graduate Size By setting (refer to Figure 12-10).

For some mapped values, you may have negative numbers. Those are left off the map by default. If you need to display negatives, select the Show Symbol option and then select the symbol for displaying them.

TIP

If you've selected a map and you want to deselect it, simply scroll SPSS Viewer to another location and select something else. When you scroll back to your map, it will be deselected.

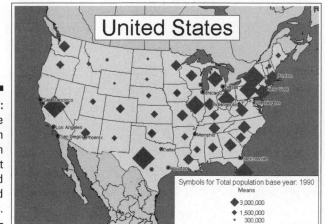

Figure 12-11:
The same population map with different and enlarged symbols.

Individual Values

Sometimes you want to present the numbers as well as the graphics of a map. You can get that by listing the individual values in a legend that keys to a map. SPSS will do this for a large map with lots of identified regions, but it is easier to read a map with fewer items.

The following steps produce an annotated population map listing the individual states of Australia:

1. **Choose File⇨Open⇨Data and open the `Australia.sav` file.**

   The file is in the `MapData` subdirectory of the SPSS directory.

2. **Choose Graphs⇨Map⇨Individual Values.**

   The Create Individual Values Map shown in Figure 12-12 appears.

3. **In the list on the left, select Total Population Current 1994, and drag it to the Individual Values For text window on the right.**

   This is the variable from which the values will be extracted.

4. **Select the State variable and drag it to the Geographic Variable text box.**

   This variable contains the location information needed for map placement.

5. **In the Geoset pull-down list, select Australia.**

   This is the map to be drawn.

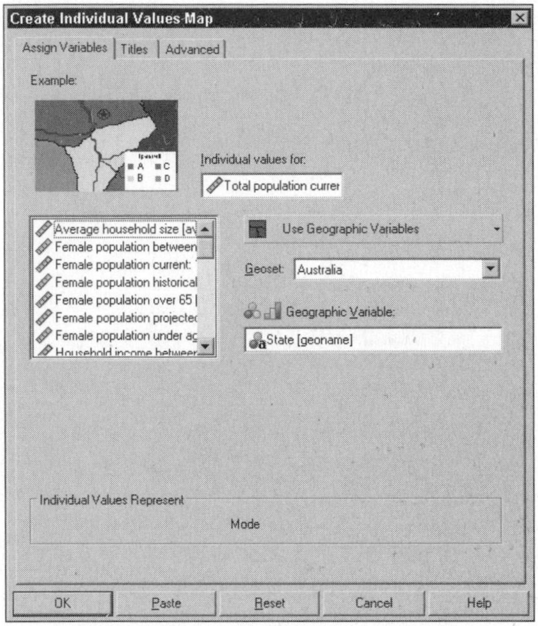

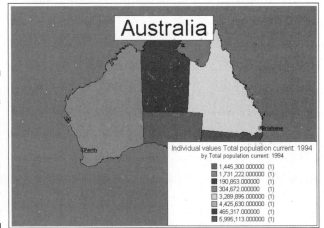

Figure 12-12:
The Create
Individual
Values Map
used to
define the
map to be
drawn.

6. **Click the OK button.**

The map shown in Figure 12-13 appears.

Figure 12-13:
A map of
Australia
with the
population
of each
state listed
in the
legend.

Bar Charts

Placing a separate bar chart in each of the geographic areas makes it possible to display the magnitude of several variables at one time for each location on the map. You can use the following steps to create such a map:

1. **Choose File⇨Open⇨Data and open the `Unites States.sav` file.**

   The file is in the `MapData` subdirectory of the SPSS directory.

2. **Choose Graphs⇨Map⇨Bar Chart.**

   The Create Bar Chart Map shown in Figure 12-14 appears.

3. **Select in turn the variables Asian, Black, Hispanic, and Caucasian and drag each one to the Bar Height text box.**

   These are the variables from which the values will be extracted.

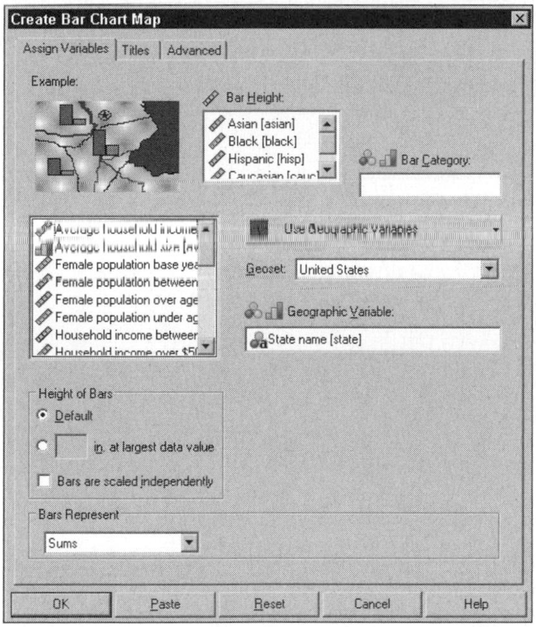

Figure 12-14: The Create Bar Chart Map used to define the map to be drawn.

4. **Select the State Name variable and drag it to the Geographic Variable text box.**

   This variable contains the location information needed for map placement.

5. **In the Geoset pull-down list, select United States.**

   This is the map to be drawn.

6. **Click the OK button.**

   A map appears. It is complete, but the bar charts on the map are too small to be seen easily. Some are no more than dots.

7. **Double-click the map to display the toolbar (refer to Figure 12-2) and then click the Themes tool (the third one from the right).**

   The Theme Control window appears.

8. **From the list of themes, select Bar Chart and then click the Display button.**

   The dialog box in Figure 12-15 appears.

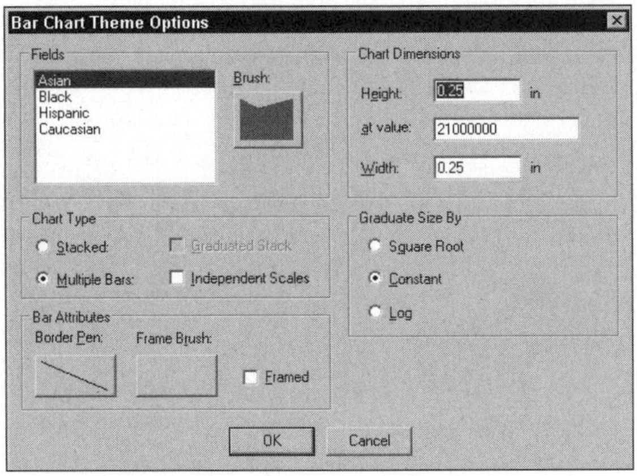

Figure 12-15: You can change the appearance of bar charts.

9. **Set the height to 1 and the width to 0.5 and then click the OK button.**

   The map looks like the one in Figure 12-16.

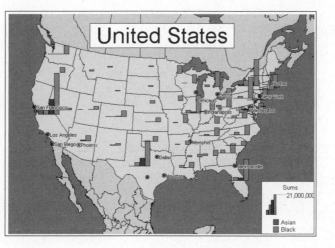

Figure 12-16:
A bar
chart map
displaying
relative
populations
by race
per state.

The Bar Chart Theme Options in Figure 12-15 can be used to modify more than just the size of the bar charts. In the upper-left corner, you can select one member of the Fields list and click the button to its right and change the color and pattern displayed in the bars for that variable.

You can select the Independent Scales option and have the size of the bars for each variable calculated in relation to the other bars of the same variable — in this example, the Independent Scales option will show you the relative numbers of each race from state to state instead of relating to the other races. You can also choose to have the bars stacked on top of one another instead of side by side.

Multiple Themes

Choose Graph⇨Maps⇨Multiple Themes to display the dialog box shown in Figure 12-17. This is the starting point for constructing a map that contains some or all of the different types of theme maps described in this chapter. If you understand how to build each individual theme map, you'll be able to build the combinations.

The tabs along the top of the dialog box can be chosen to generate information on the map. You can do this easily enough, but you need to know what you want before you start because it's easy to produce a confusing map.

Other than the capability of producing a bewildering mess, the Create Multiple Themes Maps dialog box is the same as some of the others, except the positions of some of the options have changed.

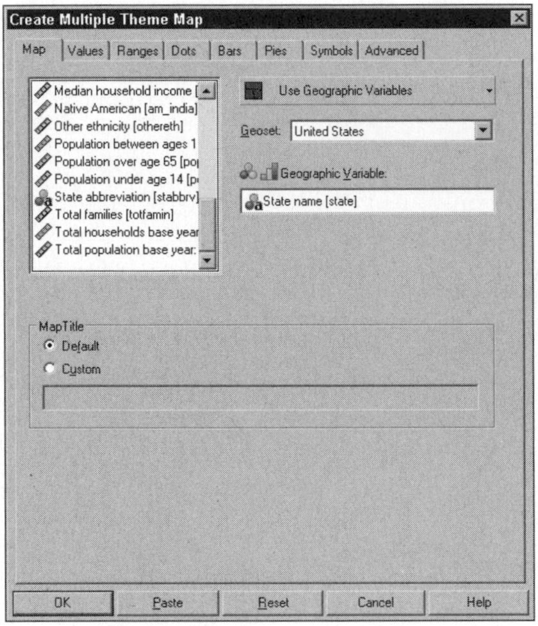

Figure 12-17:
Use this
dialog box
to construct
a combina-
tion map.

Part IV
Analysis

The 5th Wave   By Rich Tennant

"Our customer survey indicates 30% of our customers think our service is inconsistent, 40% would like a change in procedures, and 50% think it would be real cute if we all wore matching colored vests."

In this part . . .

This is the math part. SPSS is so good at it, you can almost hear the numbers crunching. But you don't need to know how to crunch the numbers — all you need to know is how to tell SPSS to crunch the numbers.

You simply select your preferred cruncher, and SPSS does the rest. Even the output is nicely formatted in a form you can use to impress others. You are Harry Potter and SPSS is your magic wand.

Chapter 13

Executing an Analysis

• •

In This Chapter

▶ Generating reports by summarizing data

▶ Displaying summary data in rows and columns

▶ Manipulating the display of pivot tables

• •

*W*hen you execute an analysis, you run your numbers through one or more processes to produce numbers that present a conclusion. In SPSS, the output from an analysis is in the form of a pivot table in SPSS Viewer. The tables are called pivot tables because you can make changes to them after they have been produced, and one of the most dramatic changes is pivoting the rows so they become columns and the columns so they become rows.

Report Generation

A report generated in SPSS is created as the result of running an analysis. The analysis can be as simple as specifying how subtotals and totals are to be calculated or as complex as the application of a multipart series of equations.

Break variables

To understand computer-generated reports, you need to understand the concept of a break variable. If a report will contain subtotals or another type of logical internal break, you must define the conditions under which the break will be made. A break usually occurs when a variable changes value. For example, if you are generating a list of employee sick days and want to insert subtotals for male and female, you could use the Gender variable as the break variable and a subtotal could be printed at the end of the 'f' values representing female and again at the end of the 'm' values representing male.

Processing summaries

When you request that SPSS create a table from your data, you also get a table labeled Processing Summary. It appears in SPSS Viewer immediately before the table you requested. Its purpose is to provide you with information about the actions taken by SPSS in the production of your table. You don't need to request a processing summary to get one.

Figure 13-1 is a simple example of a processing summary. In this example, the values from the `Engine Displacement` and `Horsepower` variables in the `Cars.sav` file were included in the table, which was organized to display information by Miles Per Gallon. In an SPSS table, the letter *N* is used as a header to indicate a simple count, or number, of items. If all the selected cases had been included in the report, there would have been 406 for each variable. In this example, a small number of cases (8 for one variable, 14 for the other) were excluded, so the report included data from 398 cases for one variable and 392 for the other. A case is excluded if the data is missing for a variable. You can see from the table that the percentage of excluded cases is quite small.

Figure 13-1:
A typical processing summary table.

Case Processing Summary

	Cases					
	Included		Excluded		Total	
	N	Percent	N	Percent	N	Percent
Engine Displacement (cu. inches) * Miles per Gallon	398	98.0%	8	2.0%	406	100.0%
Horsepower * Miles per Gallon	392	96.6%	14	3.4%	406	100.0%

Case summaries

You can construct a case summary to organize and summarize the values from one or more variables. Follow these steps:

1. **Choose File➪Open➪Data and open the `Cars.sav` file.**

   The file is in the SPSS installation directory.

2. **Choose Analyze➪Reports➪Case Summaries.**

   The Summarize Cases dialog box appears.

3. **In the list on the left:**

   a. **Select Engine Displacement and move it to the Variables panel by clicking the arrow button.**

   b. **Select Horsepower and move it to the Variables panel.**

c. Select Miles per Gallon and move it to the Grouping Variable(s) panel.

The dialog box should now look like the one in Figure 13-2, with Engine Displacement and Horsepower to be summarized, and the summaries to be grouped by Miles Per Gallon. The default, in the lower-left corner of the window, is to limit the summary to the first 100 cases and exclude cases with invalid (missing) values.

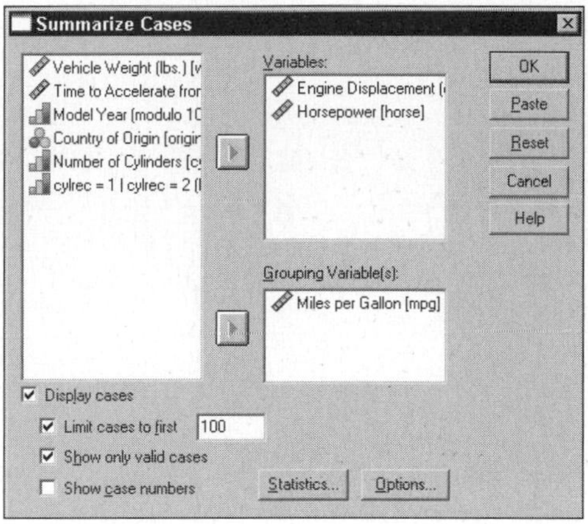

Figure 13-2: Select the variables to include in the case summary table.

4. Click the Statistics button.

The dialog box in Figure 13-3 is displayed. Here, you can select the statistics you would like to include in the report. The ones available are on the left and the ones selected are on the right.

5. Make certain the only statistic selected is Number of Cases, and then click Continue.

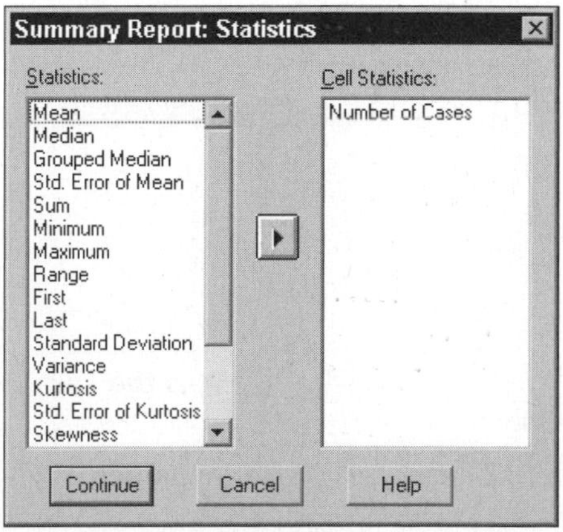

Figure 13-3: Choose the ways you want to have your summary presented.

6. Click the Options button.

The dialog box in Figure 13-4 appears. The Title is the text that appears at the top of the table, and the Caption is text that appears at the bottom. In the text you enter for either the Title or Caption, you can include \n to split the text to more than one line. You can choose whether to have missing values listed in the summary. If you do list them, it is most common to have them appear as periods or asterisks, but you can use any symbol you like.

7. Replace the default title and click Continue.

In this example, replace the default title (Case Summaries) with *Gas Mileage for Engine Size.*

8. Click Continue.

Figure 13-5 is the top portion of the table produced in this example. The entire table is not shown because it's large. The table includes data only from the first 100 cases, in which 2 cars report a gas mileage of 10 miles per hour, 2 report 11 miles per hour, and 3 report 12 miles per hour. Each car has its engine displacement and horsepower reported. The small letter *a* appended to the title indicates the presence of a footnote, which states that this report includes only the first 100 cases.

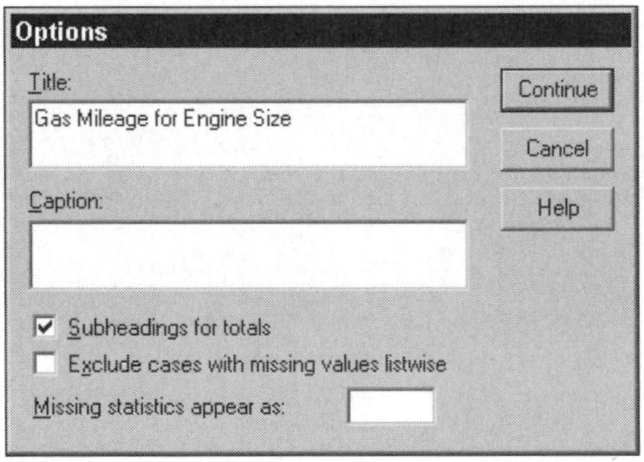

Figure 13-4:
Choose text to be placed at the top and the bottom of the table.

Gas Mileage for Engine Size[a]

				Engine Displacement (cu. inches)	Horsepower
Miles per Gallon	10	1		360	215
		2		307	200
		Total	N	2	2
	11	1		318	210
		2		429	208
		Total	N	2	2
	12	1		383	180
		2		350	160
		3		429	198
		Total	N	3	3
	13	1		400	170

Figure 13-5:
A case summary table.

Summaries in rows

You can produce a report that lists the values of a variable in a column down the left, and the values for other variables associated with it in a row to its right. Actually, you can elect to have multiple rows for each break variable by simply selecting the type of statistic.

A row summary table is simple to create but very flexible, with lots of options. This means you'll find a lot of dialog boxes, but the decisions you make are easy. Once you've run through the process a couple of times and see how it all works, you'll be able to romp through the sequence and produce output without guidance.

The following steps produce a table while giving you a tour of most of the options:

1. **Choose File⇨Open⇨Data and open the `Cars.sav` file.**

   The file is in the SPSS installation directory.

2. **Choose Analyze⇨Reports⇨Report Summaries in Rows.**

3. **In the list on the left:**

   a. **Select Engine Displacement and move it to the Data Columns panel by clicking the arrow button.**

   b. **Select Horsepower and move it to the Data Columns panel.**

   c. Select Miles per Gallon and move it to the Break Columns panel.

   The variable names in your dialog box should now look like the ones in Figure 13-6.

4. **In the Break Columns area, click the Summary button.**

   This button is enabled only if the Miles Per Gallon variable is selected. The dialog box in Figure 13-7 appears.

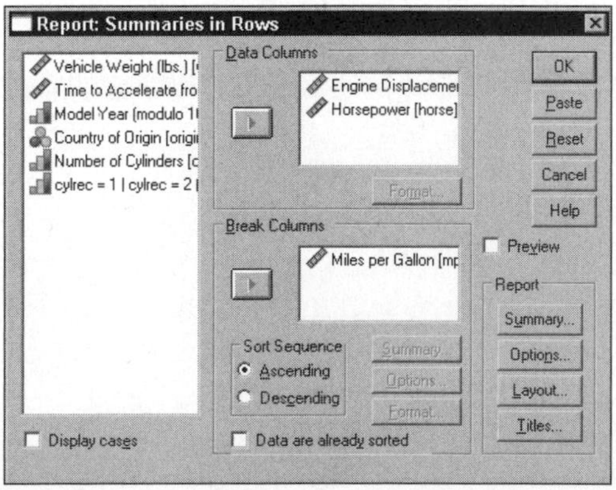

Figure 13-6: The variables selected to be included in a row summary report.

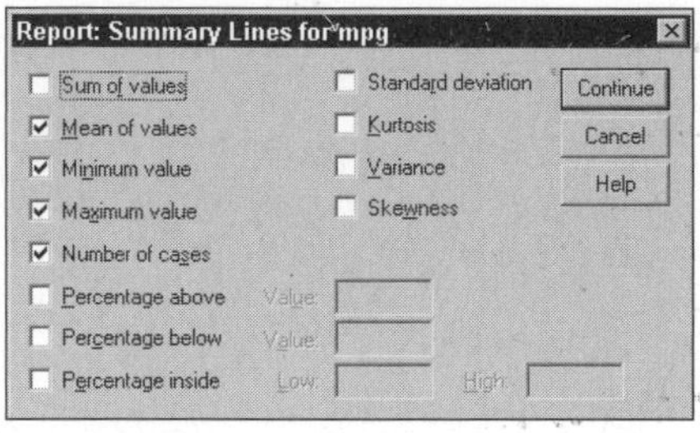

Figure 13-7: One row for each statistic type appears for each break variable value.

5. **Select the Mean of Values, Minimum Value, Maximum Value, and Number of Cases check boxes, and then click Continue.**

   A row for each of these types of statistics will be included in the report. When you click the Continue button, the dialog box closes and the dialog box shown in Figure 13-6 appears again.

6. **In the Report area, click the Summary button.**

   The dialog box in Figure 13-8 appears.

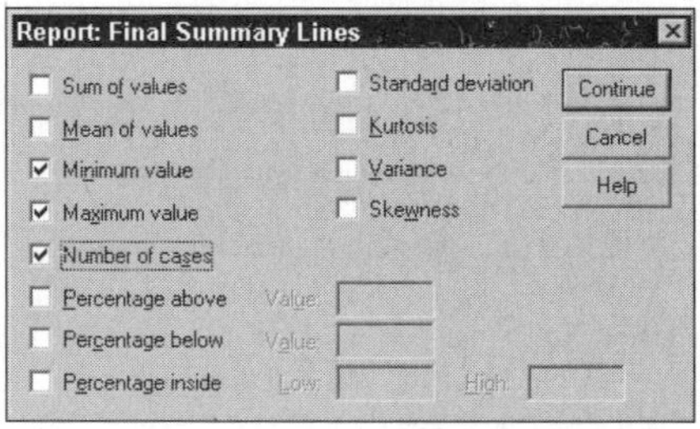

Figure 13-8: Selection of the types of summary values to appear at the bottom of the table.

7. **Select the Minimum Value, Maximum Value, and Number of Cases check boxes, and then click Continue**

   These are the values that will appear as part of the summary at the bottom of the table. (When you click the Continue button, the dialog box closes and the dialog box shown in Figure 13-6 appears again.)

8. **In the Report area, click the Options button.**

   The dialog box in Figure 13-9 appears.

9. **In the Missing Values Appear As text box, type @ (an at sign), and then click Continue.**

   The usual default in this text box is a period. You'll need to replace it with the @ sign. Missing values will be displayed as the character you enter. Alternatively, you could decide to exclude missing values entirely. (When you click the Continue button, the dialog box closes and the dialog box shown in Figure 13-6 appears again.)

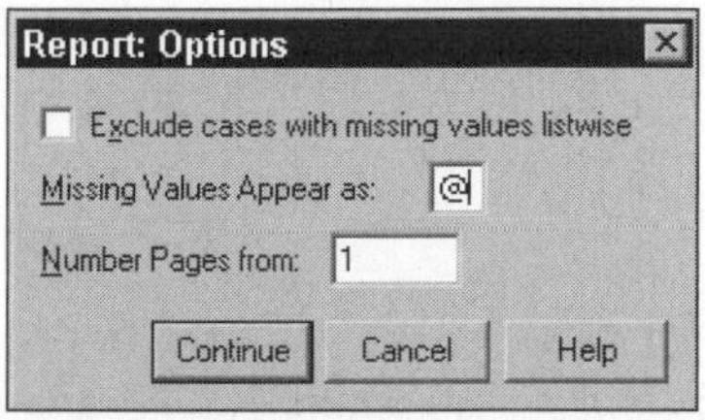

Figure 13-9:
Determine
whether
and how
missing
values are
included.

10. **In the Report area, click the Titles button.**

    The dialog box in Figure 13-10 appears.

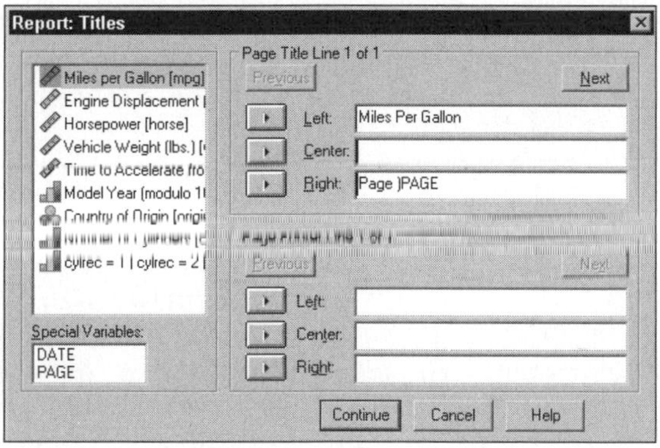

Figure 13-10:
You can
define
multiple
lines of
headers
and footers.

11. **In the upper-right text box, type the text** Miles Per Gallon, **select the Next button above it, and then enter** by Engine Size **in the text box.**

    This specifies that the heading will be two lines in length, and the text on the left will be *Miles Per Gallon by Engine Size.* The text on the right of the first line will default to the page number.

12. **Click Continue, and then click OK.**

    The output is shown in Figure 13-11. The titles are the text entered in the Titles dialog box (refer to Figure 13-10). The missing value for Miles per Gallon, displayed as @ (as specified in Step 9) occurred in 8 cases.

```
Miles Per Gallon              Page     1
by Engine Size

Miles              Engine
per             Displacement
Gallon          (cu. inches)    Horsepower

_____        _____       _____

@
Mean                    262             135
Minimum                  97              48
Maximum                 383             175
N                         8               8

10
Mean                    334             208
Minimum                 307             200
Maximum                 360             215
N                         2               2

11
Mean                    374             187
Minimum                 318             150
Maximum                 429             210
N                         4               4

12
Mean                    395             185
Minimum                 350             160
```

Figure 13-11:
A summary
in rows with
a custom
title and
missing data
displayed.

In the output, the break variable is `Miles per Gallon` and appears in the first column. Also in the first column are the names of the types of statistics, and to the right of each one is a row of values for that statistic for each variable chosen — that's why this table is known as summary in rows.

The dialog boxes in this example contain some buttons we didn't use. They all have to do with formatting details and are self-evident. You can ignore them because the defaults are reasonable, but if you want to make changes to the display, you can do so by clicking the Layout button or either Format button (refer to Figure 13-6). The Format buttons provide you with options for the display of the currently selected variable.

The action performed by the Titles dialog box (refer to Figure 13-10) may need a bit of explanation. The dialog box has two sets of three text boxes. The top set determines the text of each page's title, and the lower set determines the text of each page's footer. You can define as many lines of text for each as you want. The text boxes allow you to define the left, middle, and right of one line. As soon as you enter text for a line, the Next button becomes available and you can click it to move to the text of the next line. The Previous button allows you to back up and make changes.

Summaries in columns

You produce a report in columns by following almost the same procedure used to produce a report in rows. The options are similar, but the form of the report is quite different. You can produce a summary in column format with the following steps:

1. **Choose File⇨Open⇨Data and open the `Home sales [by neighborhood].sav` file.**

   The file is in the SPSS installation directory.

2. **Choose Analyze⇨Reports⇨Report Summaries in Columns.**

   The Summarize Cases dialog box appears.

3. **In the list on the left:**

   a. **Choose Appraised Land Value and move it to the Data Columns panel by clicking the arrow button.**

      It appears with its name and statistic type as landval:sum.

   b. **Select Appraised Value of Improvements and move it to the Data Columns panel.**

      Its name and statistic type appear as improval:sum.

   c. **Select Neighborhood and move it to the Break Columns panel.**

4. **Click the Insert Total button.**

   The word Total (defining a new column) will be added to the bottom of the list in the Data Columns list. Your dialog box should now look like the one in Figure 13-12.

5. **Select landval:sum from the list and then click Summary.**

   The dialog box in Figure 13-13 appears.

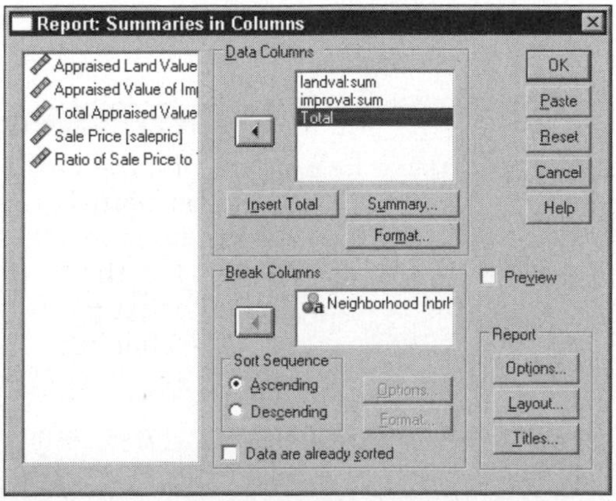

Figure 13-12: The variables that will appear in the summaries in columns report.

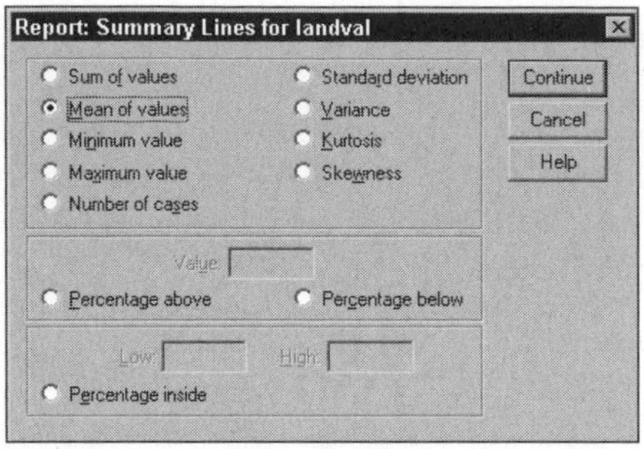

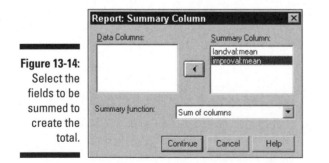

Figure 13-13:
Specify
which
statistic
value will be
calculated
for a
variable.

Report: Summary Lines for landval

- ○ Sum of values
- ⦿ Mean of values
- ○ Minimum value
- ○ Maximum value
- ○ Number of cases
- ○ Standard deviation
- ○ Variance
- ○ Kurtosis
- ○ Skewness

Continue
Cancel
Help

Value:
○ Percentage above   ○ Percentage below
Low: High:
○ Percentage inside

6. **Select Mean of Values and then click Continue.**

   The first variable in the Data Columns panel is now listed as landval:mean to show that the variable is the same as before, but the statistic is now mean instead of sum.

7. **Select Total in the Data Columns panel and then click Summary.**

   The Summary Column dialog box appears.

8. **Select landval:mean in the Data Columns panel and click the arrow button to move it to the Summary Column panel (see Figure 13-14). Do the same for improval:mean.**

   You are choosing the variables to be summed to produce the total. You could calculate the total in ways other than a simple sum by selecting another option from the pull-down list, but the default Sum of Columns is right for this example.

Figure 13-14:
Select the
fields to be
summed to
create the
total.

Report: Summary Column

Data Columns:

Summary Column:
landval:mean
improval:mean

◀

Summary function: Sum of columns ▼

Continue   Cancel   Help

9. **Click Continue and then click OK.**

   The table is output and displayed by SPSS Viewer, as shown in Figure 13-15.

For each neighborhood listed in the first column, the report shows the mean land appraisal value, the mean appraisal value of the improvements, and the total of the two means — the total mean appraisal value.

```
                                        Page      1

              Appraised    Appraised
                Land        Value of
                Value     Improvements
Neighborhood     Mean         Mean        Total

A               44718        47064        91782

B               26671        54302        80974

C               24175        48461        72636

D               18221        61452        79673

E               13729        41048        54777

F               14297        25426        39722

G                8847        19731        28579
▽
```

Figure 13-15:
A summary
in columns
report with
a total
column.

Other options are available for defining the appearance of this report, but the defaults are reasonable and probably should be used unless you have something specific in mind. The Titles button allows you to specify the text of the headers and footers using the same technique as that used in the summaries in rows report.

OLAP cubes

A regular table is in two dimensions: height and width. A cubed table is in three dimensions: height, width, and depth. It's like a deck of cards with a regular two-dimensional table printed on each card. You can flip from one card to another to see any of the tables. Thus it adds the third dimension, depth, and becomes cubed.

An _OLAP (Online Analytical Processing) cube_ is the output of a process that uses one or more scale variables along with one or more categorical values to divide the report information into layers for the depth. The following steps guide you through the process of producing a three-dimensional table:

1. **Choose File➪Open➪Data and open the Employee data.sav file.**

   The file is in the SPSS installation directory.

2. **Choose Analyze➪Reports➪OLAP Cubes.**

   The OLAP Cubes dialog box appears, as shown in Figure 13-16.

3. **In the list on the left:**

   a. Select Current Salary and move it to the Summary Variable(s) panel by clicking the arrow button.

   b. **Select Beginning Salary and move it to the Summary Variable(s) panel.**

   c. **Select Educational Level and move it to the Grouping Variable(s) panel.**

   d. **Select Employment Category and move it to the Grouping Variable(s) panel.**

   The results should look like Figure 13-17. This will produce a table with several layers — the beginning salary and the current salary will each be shown in separate tables based on educational level and job category. The Statistics button is now available in the dialog box because variables to make up a valid table have been chosen.

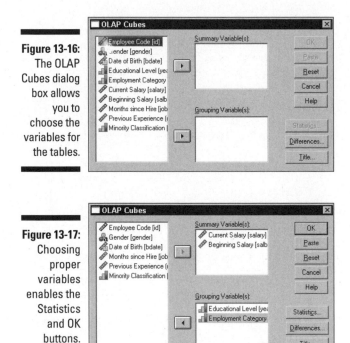

Figure 13-16: The OLAP Cubes dialog box allows you to choose the variables for the tables.

Figure 13-17: Choosing proper variables enables the Statistics and OK buttons.

4. **Click the Statistics button.**

   The OLAP Cubes Statistics dialog box appears, as shown in Figure 13-18. In this dialog box, you decide what calculations you want SPSS to perform.

5. **Change the list of selected Cell Statistics to include only Number of Cases, Minimum, Maximum, Kurtosis, Skewness, and Grouped Median.**

These are the statistics that will be calculated for the table. To select a statistic, highlight its name in the list on the left and click the arrow button to move it to the right. To deselect a statistic, select its name in the list on the right and click the arrow button.

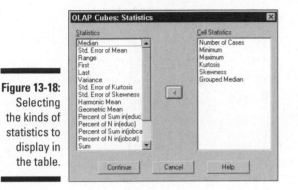

Figure 13-18:
Selecting
the kinds of
statistics to
display in
the table.

The order in which the values appear in the table is determined by the order in which their names appear in the list — you determine the order by moving them into the list in the order you want them to appear. To change the order, you can take them out and then move them back in the order you want.

6. **Click the OK button.**

   The table in Figure 13-19 appears. This is only the total layer of the multilayered table.

Figure 13-19:
One layer
of a multi-
layered
table
displaying
the statistics
for the total.

OLAP Cubes

Educational Level (years): Total
Employment Category: Total

	N	Minimum	Maximum	Kurtosis	Skewness	Grouped Median
Current Salary	474	$15,750	$135,000	5.378	2.125	$28,900.00
Beginning Salary	474	$9,000	$79,980	12.390	2.853	$14,988.68

Double-clicking the OLAP Cubes table selects it and causes the appearance of pull-down lists, as shown in Figure 13-20. One pull-down list appears for each grouping variable. By making selections from the lists, you change the view by changing the table that appears on top.

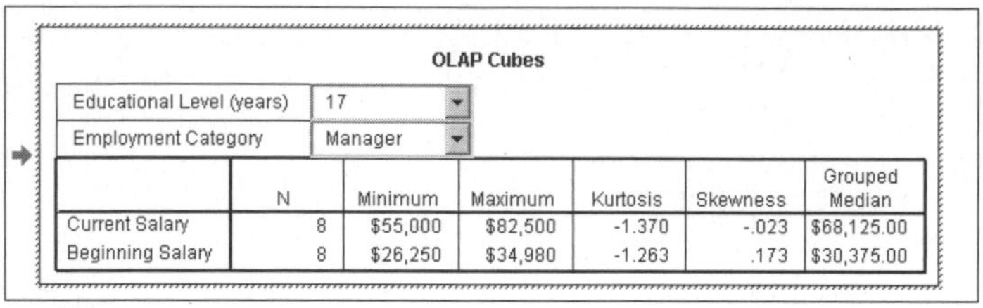

Figure 13-20:
An OLAP
cube stack
of tables
with two
variables
determining
which table
is in view.

Pivot Tables

The tables that appear as output in SPSS Viewer are called *pivot tables* because you can change their appearance in several ways — not the least of which is to pivot the table by swapping the rows and columns.

To make modifications to a table in SPSS Viewer, you first select a table and then choose View⇨Toolbar. A toolbar like the one in Figure 13-21 appears. You can use the toolbar to modify the font and alignment of the text in the table.

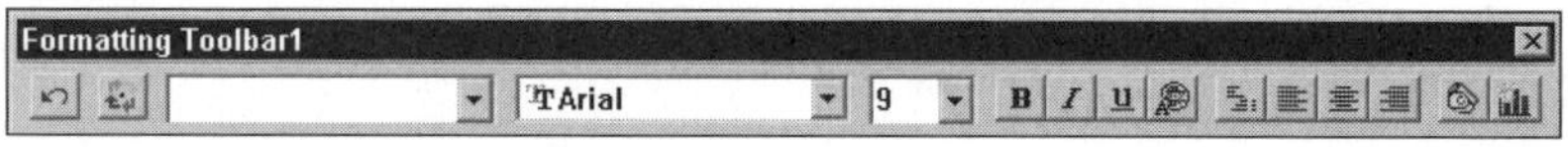

Figure 13-21:
Use the
toolbar to
modify the
table's
appearance.

To perform a table pivot, click the button that's second from the left on the toolbar. The dialog box shown in Figure 13-22 appears.

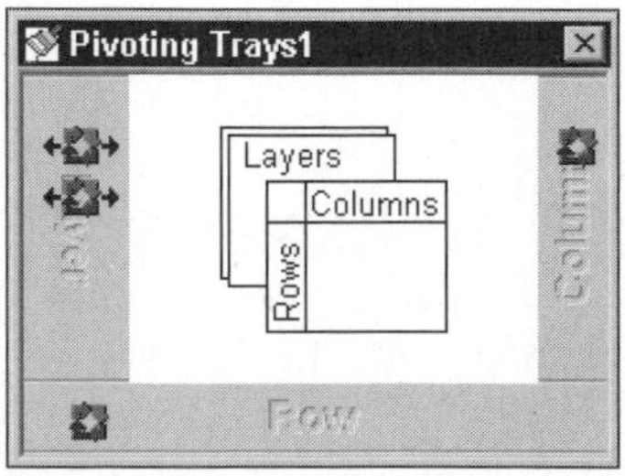

Figure 13-22:
The posi-
tions of
variables in
the table.

As you change the position of items in the dialog box, the display of the table in SPSS Viewer changes to the new configuration:

- ✔ Each black square with the diamond shape in its middle is a variable displayed in the table.

- ✔ Squares on the right of the dialog box represent variables in columns.

- ✔ Those on the bottom represent variables in rows.

- ✔ Those on the left (with arrows pointing out the side) are layers in a multilayered table.

Dragging a black square from one location to another in the dialog box reshapes the table with that variable in the new location. By dragging, it's easy to move a column variable so it becomes a row variable. You can even drag row and column variables to make them layered variables, and vice versa. The table can be reshaped dramatically.

You can switch the current view of a layered table by clicking the arrows on the variable symbols on the left. This has the same effect as selecting values from the pull-down lists on a multilayered table.

Chapter 14

Some Analysis Examples

his chapter describes how to instruct SPSS to dig into your data, execute an analysis, and reach a conclusion. In SPSS, executing an analysis involves taking your raw data, performing calculations on it, and presenting the results in a table or a chart.

This chapter provides examples of the most fundamental types of analysis that SPSS offers. Menu choices and options that I don't demonstrate are more advanced forms of the same types of analysis and require more input — a slightly different kind of input — but they employ the same basic algorithms. In general, an understanding of the way the analysis examples in this chapter operate will give you the understanding you need for the more advanced forms of analysis.

In the descriptions in this chapter, I assume that you're familiar with the fundamental procedures required for constructing tables, which I describe in Chapter 13.

Comparison of Means

The tests for comparing the mean of one variable to the mean of another are more varied and flexible than you might think. The analysis methods in this section fall into the category of means tests, but they are actually more than that. You'll find that they can produce up to twelve statistics, of which the mean is only one.

Simple mean compare

You can generate a simple comparison table by loading the `Employee data.sav` file and choosing Analyze⇨Compare Means⇨Means. The dialog box in Figure 14-1 appears, with a list of variable names on the left. Select the variables to be used for calculating the mean — Beginning Salary and Current Salary — and transfer them to the Dependent List panel (by clicking the arrow button). Select the Employment Category variable and move it to the Independent List panel. This is all you have to do to produce output.

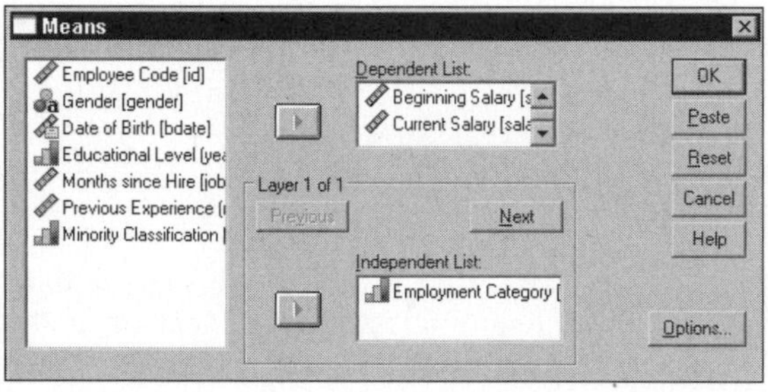

Figure 14-1: Choosing the variables that will generate the table.

The table produced from this dialog box can include more than simply the mean. By clicking Options, you can choose from a combination of 21 statistics. The default selections are Mean, Number of Cases, and Standard Deviation. Using the default statistics and the variables selected in Figure 14-1, I generated the table shown in Figure 14-2.

Figure 14-2: Comparison of means and standard deviation according to employment category.

Report			
Employment Category		Beginning Salary	Current Salary
Clerical	Mean	$14,096.05	$27,838.54
	N	363	363
	Std. Deviation	$2,907.474	$7,567.995
Custodial	Mean	$15,077.78	$30,938.89
	N	27	27
	Std. Deviation	$1,341.235	$2,114.616
Manager	Mean	$30,257.86	$63,977.80
	N	84	84
	Std. Deviation	$9,980.979	$18,244.776
Total	Mean	$17,016.09	$34,419.57
	N	474	474
	Std. Deviation	$7,870.638	$17,075.661

You can include other independent variables in two ways. The table in Figure 14-2 is single layered, but by clicking the Next button in the dialog box in Figure 14-1, you can add new layers for independent variables. You can also add independent variables to the same top layer (or any other layer) and make the table larger to include them.

One-sample T test

The *one-sample T test* analysis compares an expected value with the mean derived from the values of a single variable. To run the test, you choose the variable to be averaged and the value you expect. The report shows you the accuracy of your expectations.

For an example of the T test, open the `Employee data.sav` file. Choose Analyze⇨Compare Means⇨One Sample T Test and the dialog box in Figure 14-3 appears. As shown in Figure 14-3, I selected the Educational Level variable and the number 12. The mean of the variable will be compared against the constant value 12.

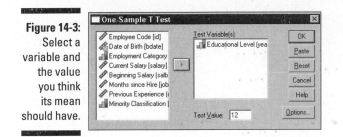

Figure 14-3: Select a variable and the value you think its mean should have.

The resulting table is shown in Figure 14-4. At the top of the table is the value that's the basis of all comparisons — the average number of years of educa-tion of all employees was compared to 12. The first column, labeled with the letter *t,* is the mean value derived from the data. The second column, the one labeled *df,* is the degrees of freedom. The Mean Difference column is the aver-age of the magnitude of the differences of the values from the expected value. The Confidence Interval values show how wide the range is around the value of 12 to include 95 percent of all values.

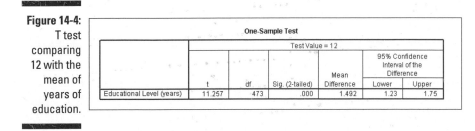

Figure 14-4: T test comparing 12 with the mean of years of education.

One-Sample Test

	Test Value = 12				95% Confidence Interval of the Difference	
	t	df	Sig. (2-tailed)	Mean Difference	Lower	Upper
Educational Level (years)	11.257	473	.000	1.492	1.23	1.75

Independent-samples T test

The *independent-samples T test* compares the means of two sets of values from one variable. To run an example of the test, load the `Employee data.sav` file. Choose Analyze➪Compare Means➪Independent-Samples T Test, and the dialog box in Figure 14-5 appears.

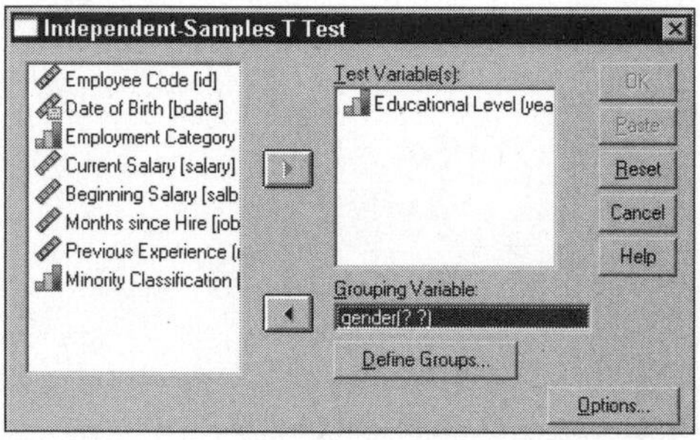

Figure 14-5:
Test to compare the means of two variables.

Move the Educational Level variable to the Test Variable(s) panel. This variable will supply the values for the means to be tested. Move the Gender variable to the Grouping Variable panel; this is the variable that will be used to select the two groups. The variable could have multiple values defined for it, but you need to choose only two. Click the Define Groups button to specify the two values — in this example, the only values available are `m` and `f`. Entering these two values causes them to appear in place of the question marks following the name of the variable. Click the OK button, and the pair of tables in Figure 14-6 is produced.

Group Statistics

	Gender	N	Mean	Std. Deviation	Std. Error Mean
Educational Level (years)	Male	258	14.43	2.979	.185
	Female	216	12.37	2.319	.158

Figure 14-6:
The pair of tables produced from the independent-samples T test.

Independent Samples Test

		Levene's Test for Equality of Variances		t-test for Equality of Means						
Dependent variables	Assumptions	F	Sig.	t	df	Sig. (2-tailed)	Mean Difference	Std. Error Difference	95% Confidence Interval of the Difference	
		Lower	Upper	Lower	Upper	Lower	Upper	Lower	Upper	Lower
Educational Level (years)	Equal variances assumed	17.884	.000	8.276	472	.000	2.060	.249	1.571	2.549
	Equal variances not assumed			8.458	469.595	.000	2.060	.244	1.581	2.538

The table displays the two means and the standard deviation and standard error for the two means. The Independent Samples Test table provides further information about the mean in two rows of numbers — one for equal variances and one for unequal variances:

- ✔ If the significance of the Levene test, the number in the second column, is high (greater than 0.05 or so), the values in the first row are applicable.

- ✔ If the significance of the Levene test is low, the numbers in the second row are more applicable.

- ✔ If the significance of the T test, the 2-tailed significance, is low, this indicates a significant difference in the two means.

- ✔ If none of the numbers of the 95% confidence interval are 0, it indicates the difference is significant.

Paired-samples T test

The *paired-samples T test* is a comparison test specially designed to compare values from the same group at different times. The values could be gathered before and after an event, or before and after a passage of time.

To run the test, choose Analyze⇨Compare Means⇨Paired-Samples T Test. You select two variable names from the list on the left, click the arrow button, and the two show up as a pair on the right, as shown in Figure 14-7. That's all there is to it unless you want to use the Options button to change the 95% confidence level to another percentage. Click the OK button to produce the paired-samples T test table.

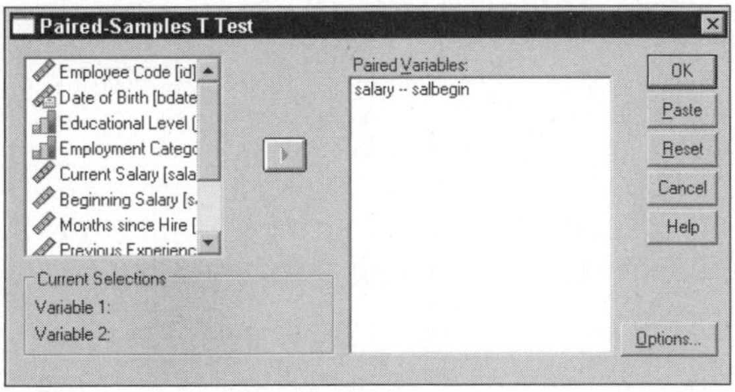

Figure 14-7: Selecting two variables causes them to appear on one line in the panel

One-way ANOVA

ANOVA is an analysis of variance. A one-way ANOVA is the analysis of the variance of the values (of a dependent variable) by comparing them against another set of values (the independent variable). It is a test of the hypothesis that the mean of the tested variable is equal to that of the factor.

The output table from running this test is a small one. To see an example of its output, load the `Road construction bids.sav` file. Then choose Analyze⇨Compare Means⇨One-Way ANOVA. In the dialog box shown in Figure 14-8, I'm testing the hypothesis that the mean of the contractor's construction costs matches that of the department of transportation's engineering cost estimates. The result is the table shown in Figure 14-9.

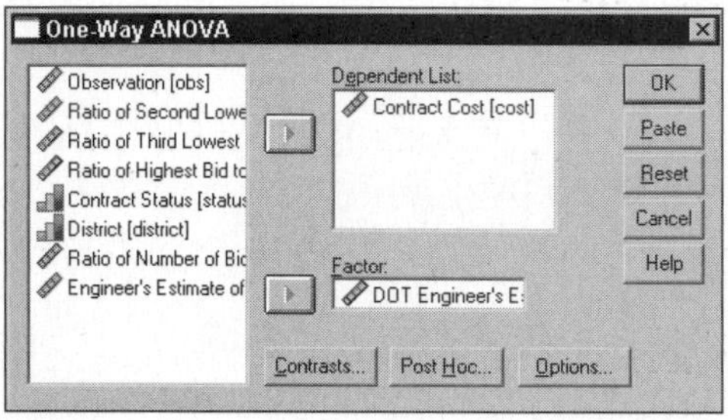

Figure 14-8:
One variable is chosen to be tested and another is chosen as the factor to test against.

Figure 14-9:
The analysis of the variance of one variable as compared to that of another.

ANOVA

Contract Cost

	Sum of Squares	df	Mean Square	F	Sig.
Between Groups	8.9E+008	233	3806325.945	3704.798	.013
Within Groups	1027.404	1	1027.404		
Total	8.9E+008	234			

Linear model

Many statistical values result from comparing actual results against expected results — or, in statistics speak, the comparison of dependent variables against independent variables. Straight lines are easier to compare than curves and often produce a result that's easier to understand. This section is about curveless analysis.

One variable

You can compare one dependent variable against more than one independent variable. For example, suppose a plastic manufacturer wants to increase the tear resistance of his product, so he varies the extrusion rate and additives

to do so. To see how the results of the study can be calculated, open the
`Plastic.sav` file. Then choose Analyze⇨General Linear Model⇨Univariate.

The Tear Resistance variable is selected to be the one dependent variable,
and the two variables Additive Amount and Extrusion are chosen as the fixed
variables, as shown in Figure 14-10.

The table in Figure 14-11 is produced, displaying the resulting values of Tear
Resistance depending on Extrusion and Additive Amount, both individually
and together.

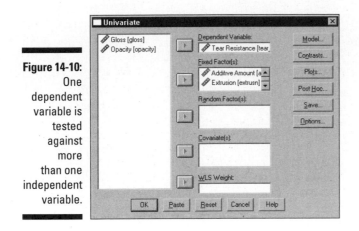

Figure 14-10:
One
dependent
variable is
tested
against
more
than one
independent
variable.

Tests of Between-Subjects Effects

Dependent Variable: Tear Resistance

Source	Type III Sum of Squares	df	Mean Square	F	Sig.
Corrected Model	2.501 [a]	3	.834	7.563	.002
Intercept	920.724	1	920.724	8351.243	.000
additive	.760	1	.760	6.898	.018
extrusn	1.740	1	1.740	15.787	.001
additive * extrusn	.000	1	.000	.005	.947
Error	1.764	16	.110		
Total	924.990	20			
Corrected Total	4.265	19			

a. R Squared = .586 (Adjusted R Squared = .509)

Figure 14-11:
The Tear
Resistance
variable is
tested
against the
effect of two
factors.

More than one variable

It is also possible to measure more than one dependent variable against
more than one independent variable. Using the same data as in the single-
value test of the preceding section, choose Analyze⇨General Linear Model⇨
Multivariate. The Gloss, Tear Resistance, and Opacity dependent variables
will be tested against the Additive Amount and Extrusion variables, as shown
in Figure 14-12.

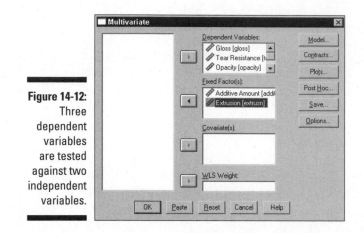

Figure 14-12:
Three
dependent
variables
are tested
against two
independent
variables.

Click the OK button, and the table in Figure 14-13 is produced. You may notice that this table is the same basic form as the single-value table in the preceding section, except the Dependent Variable column now has three entries for each entry in the Source column.

Tests of Between-Subjects Effects

Source	Dependent Variable	Type III Sum of Squares	df	Mean Square	F	Sig.
Corrected Model	Gloss	2.457[a]	3	.819	4.987	.012
	Tear Resistance	2.501[b]	3	.834	7.563	.002
	Opacity	9.282[c]	3	3.094	.762	.531
Intercept	Gloss	1735.385	1	1735.385	10565.507	.000
	Tear Resistance	920.724	1	920.724	8351.243	.000
	Opacity	309.684	1	309.684	76.319	.000
additive	Gloss	.612	1	.612	3.729	.071
	Tear Resistance	.760	1	.760	6.898	.018
	Opacity	4.900	1	4.900	1.208	.288
extrusn	Gloss	1.301	1	1.301	7.918	.012
	Tear Resistance	1.740	1	1.740	15.787	.001
	Opacity	.420	1	.420	.104	.752
additive * extrusn	Gloss	.544	1	.544	3.315	.087
	Tear Resistance	.000	1	.000	.005	.947
	Opacity	3.960	1	3.960	.976	.338
Error	Gloss	2.628	16	.164		
	Tear Resistance	1.764	16	.110		
	Opacity	64.924	16	4.058		
Total	Gloss	1740.470	20			
	Tear Resistance	924.990	20			
	Opacity	383.890	20			
Corrected Total	Gloss	5.085	19			
	Tear Resistance	4.265	19			
	Opacity	74.206	19			

a. R Squared = .483 (Adjusted R Squared = .386)

Figure 14-13:
Tear
Resistance,
Gloss, and
Opacity all
being tested
against the
effect of two
factors.

Correlation

The group of tests in this section determines the similarity or difference in the way two variables change in value from one case (row) to another through the data.

Bivariate

To run a simple bivariate (two-variable) correlation, load data that has two variables to be compared and choose Analyze⇨Correlate⇨Bivariate. In Figure 14-14, I'm performing a test to determine whether there's a correlation between an employee's starting salary and current salary.

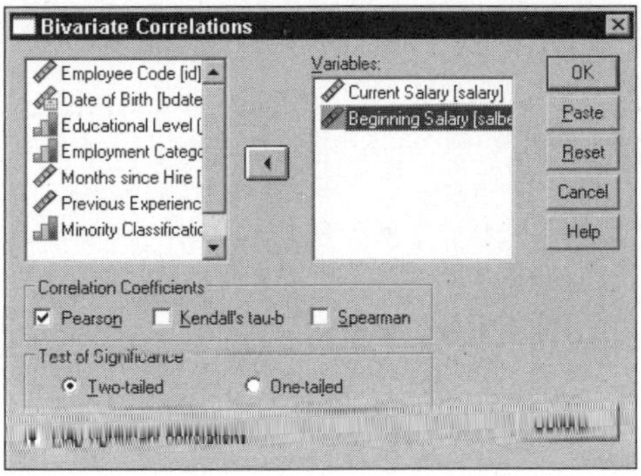

Figure 14-14:
Select variables to be compared by moving them to the right.

You can choose up to three kinds of correlations. The most common form is the Pearson correlation, which is the default. If you want, you can click the Options button and decide what to do about missing values and tell SPSS whether you want to calculate the standard deviations. The result of the selections in Figure 14-14 is shown in Figure 14-15.

Figure 14-15:
Pearson correlation showing a highly significant correlation.

Correlations

		Current Salary	Beginning Salary
Current Salary	Pearson Correlation	1	.880**
	Sig. (2-tailed)		.000
	N	474	474
Beginning Salary	Pearson Correlation	.880**	1
	Sig. (2-tailed)	.000	
	N	474	474

**. Correlation is significant at the 0.01 level (2-tailed).

Correlation figures vary from –1 to +1, and the larger the value, the stronger the correlation. In Figure 14-15, you can see that the variables have a correlation of 1 with themselves and .880 with one another, which is a significant correlation.

Partial correlation

Outside factors can affect a correlation. You can include these factors in the calculations; such a test is known as a *partial correlation*. For example, in the previous example, I found that the current salary of each employee correlated with the starting salary, but I did not take into account the length of employment. In this example, I will. Begin by choosing Analyze⇨Correlate⇨ Partial.

Select the Current Salary and Beginning Salary, along with the Months Since Hire as the factor that should have an effect on the correlation. The dialog box should look like the one in Figure 14-16.

The result is an even higher level of correlation than before, as shown in Figure 14-17.

Figure 14-16: Select the variables to correlate and the variable to control the correlation.

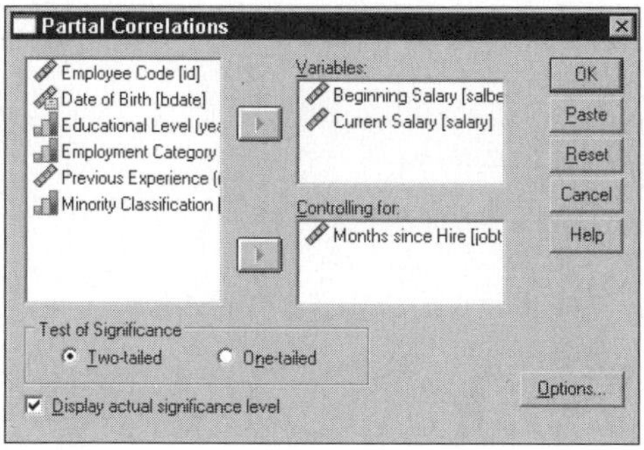

Figure 14-17: The correlation of starting with the current salary and taking the length of employment into account.

Correlations

Control Variables			Beginning Salary	Current Salary
Months since Hire	Beginning Salary	Correlation	1.000	.885
		Significance (2-tailed)	.	.000
		df	0	471
	Current Salary	Correlation	.885	1.000
		Significance (2-tailed)	.000	.
		df	471	0

Regression

Regression analysis is about predicting the future (the unknown) based on data collected from the past (the known). A *regression analysis* determines the mathematical equation to be used to figure out what will happen, within a certain range of probability. It analyzes one variable, the dependent variable, taking into consideration the effect on it by one or more factors, the independent variables. The analysis determines that some independent variables have more effect than others, so their weights must be taken into account when they are the basis of a prediction. Regression analysis, therefore, is the process of looking for predictors and determining how well they predict.

When only one independent variable is taken into account, it's called a simple regression. If you use more than one independent variable, it's called multiple regression. All the dialog boxes of SPSS provide for multiple regression.

Linear

Linear regression is used when the projections are expected to be in a straight line with actual values. The following is an example of a linear multiple regression:

1. **Choose File⇨Open⇨Data and open the `sales.sav` file.**

   The file is in the SPSS installation directory.

2. **Choose Analyze⇨Regression⇨Linear.**

   The Summarize Cases dialog box appears.

3. **Select Revenue and move it to the Dependent panel.**

   This is the variable for which we want to set up a prediction equation.

4. **Select the other four variables and move them to the Independent(s) panel.**

   The screen should look like Figure 14-18. The resulting equation will include Customer Status, Time on Hold, Territory, and Industry. The assumption is made that all four have an effect on the size of the revenue.

5. **Click OK.**

   The table in Figure 14-19 is produced.

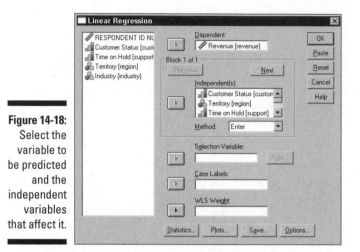

Figure 14-18:
Select the variable to be predicted and the independent variables that affect it.

Figure 14-19:
The table containing coefficients for making revenue predictions.

Coefficients[a]

Model		Unstandardized Coefficients		Standardized Coefficients		
		B	Std. Error	Beta	t	Sig.
1	(Constant)	3248.991	130.121		24.969	.000
	Customer Status	-91.509	50.106	-.046	-1.826	.068
	Territory	28.129	21.813	.032	1.290	.197
	Time on Hold	-242.171	21.723	-.278	-11.148	.000
	Industry	4.565	30.637	.004	.149	.882
a. Dependent Variable: Revenue						

You will find other tables included as part of the output, but they all have to do with how the values of this table are produced. This table defines the equation for you in the first column. Revenue can be predicted with the following:

```
Revenue = 3248.991 - (91.509)(Customer Status) +
          (28.129)(Territory) - (242.171)(Time On Hold) +
          (4.565)(Industry)
```

Curve estimation

If you have a collection of data points, it's possible to create a curve that passes through (or very near) those points. That curve can then be used to estimate the values of points you don't have yet. This can be done by *interpolation* (drawing a curve connecting the existing points) or *extrapolation* (extending the curve beyond the existing points). The graphic presentation of values isn't as numerically accurate as a table of numbers, but it has some advantages, not least of which is the ability to quickly spot patterns and trends. Predictions are only estimations no matter how sophisticated, so presenting a prediction as a graph is as good as with numbers even with the inherent inexactness.

In the following, I fit a curve to a group of data points for the purpose of demonstrating the probable horsepower of an engine depending on its cubic inches of displacement:

1. **Choose File⇨Open⇨Data and open the Cars.sav file.**

   The file is in the SPSS installation directory.

2. **Choose Analyze⇨Regression⇨Curve Estimation.**

   The Curve Estimation dialog box appears.

3. **Select Horsepower as the variable to have its value predicted by moving it to the Dependent(s) panel.**

   You could choose more than one dependent variable and the output would be more than one chart. Each dependent variable has its own graph.

4. **Select Engine Displacement and move it to the Independent panel.**

5. **Select Linear, Quadratic, and Cubic as the types of curves to be generated.**

   The screen should look like Figure 14-20.

6. **Click OK.**

   Some tables are generated to describe the processing SPSS used to reach its conclusion. The graph shown in Figure 14-21 contains the three requested curves.

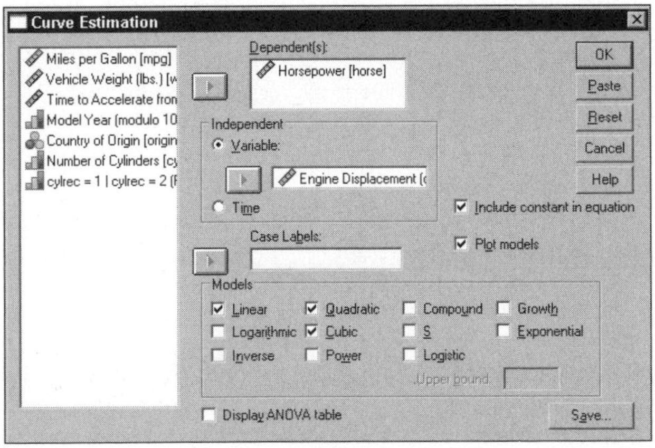

Figure 14-20: Select the variables involved in curve fitting and the types of curves.

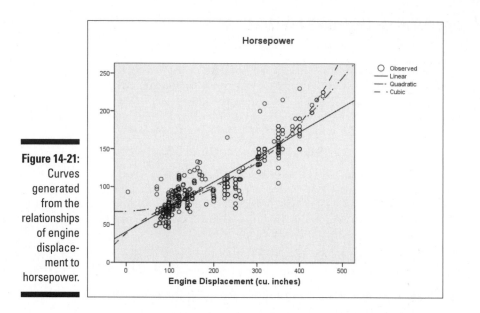

Figure 14-21:
Curves
generated
from the
relationships
of engine
displace-
ment to
horsepower.

In Figure 14-21, each dot represents the relationship of actual engine displacement to measured horsepower. The predicted values of horsepower according
to displacement are represented in three ways. The linear interpretation is the
best fit of a straight line to the dots. The quadratic line is the best fit of a line
that curves in one direction. The cubic line reverses the direction of its curve
in an attempt to fit as closely as possible. None of the curves fit the data
points exactly, but they give you the best possible prediction of the results.

Log Linear

Log linear is based on the assumption that a linear relationship exists between
the independent variables and the logarithm of the dependent variable.

The example in this section summarizes the expected starting salaries of
college graduates, organizing the summaries by gender and the college
from which they graduated. To generate this table, open the `graduate`
`salaries.sav` file. Then choose Analyze⇨Loglinear⇨General.

Move the Gender and College variables to the Factor(s) panel, making them
the two variables used to divvy up the results. Move the Starting Salary variable to the Contrast Variable(s) panel, making it the variable containing the
data to be divvied up. Your screen should look like Figure 14-22.

Click the OK button and the table shown in Figure 14-23 appears. You can
see that the salaries for engineering are high for both genders. In this table
of salaries, there is no clear difference according to sex.

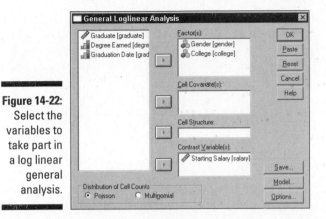

Figure 14-22:
Select the variables to take part in a log linear general analysis.

Coefficients[b,c]

Gender	College	Starting Salary[a]
Female	Agriculture	24199
	Architecture	21000
	Building/Construction	29825
	Business Administration	23969
	Forestry	14500
	Education	21875
	Engineering	31378
	Fine Arts	20900
Male	Agriculture	22992
	Architecture	22150
	Building/Construction	28033
	Business Administration	25409
	Forestry	23500
	Education	19800
	Engineering	30781
	Fine Arts	26000

a. Sum of the coefficients is not zero. The generalized log-odds ratio is not computed.

b. Model: Poisson

c. Design: Constant + gender + college + gender * college

Figure 14-23:
Starting salaries divided by sex and further divided by the type of degree.

Executing the same analysis but leaving out the variable for the type of degree, we get a table that organizes salaries only by sex, as shown in Figure 14-24.

Figure 14-24:
Starting salaries grouped by sex and no other factors.

Coefficients[b,c]

Gender	Starting Salary[a]
Female	24770
Male	27027

a. Sum of the coefficients is not zero. The generalized log-odds ratio is not computed.

b. Model: Poisson

c. Design: Constant + gender

Part V

Programming SPSS with Command Syntax

The 5th Wave   By Rich Tennant

"It's an electronic wedding planner. It'll create all your check lists and time tables, and after the ceremony it turns all the documents into confetti and throws it in your face."

In this part . . .

Down inside SPSS, where its heart beats, everything happens because of statements written in the Command Syntax language. You can skip the menus and dialog boxes and issue commands directly to the internals of SPSS. It may sound a bit spooky at first, but it isn't as hard as it sounds. In fact, SPSS will help you write Command Syntax statements.

Chapter 15

The Command Syntax Language

*E*verything that happens in SPSS is the result of executing a Command Syntax script. Whenever you use the menu to specify a set of options and then click an OK button instructing SPSS to perform some feat, a Command Syntax script is generated and put into execution. This chapter and the next are about the language called Command Syntax, or Syntax for short.

Commands

A single Syntax language instruction can be very simple or complex enough to serve as an entire program. A single instruction consists of a command followed by arguments to modify or expand the actions of the command. For example, the following Syntax command generates a report:

```
REPORT /FORMAT=LIST /VARIABLES=MPG.
```

The first thing you probably noticed is that the command is written in all uppercase. That's tradition — not a requirement. You can write in lowercase (or even mixed case) if you want. Notice also that the end of the list of arguments is terminated by a single period. The terminator must be there or SPSS will complain.

Now, about those forward slashes and equal signs. Sometimes you need them, and sometimes they're optional. Always use them and you won't have any trouble. The presence of slashes and equal signs reduces ambiguity for you and SPSS. Also, commands can be abbreviated as long as you have at least three letters to uniquely identify each command. I can't think of a single reason to abbreviate anything. Figuring out how to abbreviate a command is more work than just typing it, and abbreviation makes the program harder to read.

The command in this example is REPORT, which causes text to be written to SPSS Viewer. In fact, all output produced by running Syntax programs goes to SPSS Viewer. The FORMAT specification tells REPORT to make a list of the values. The VARIABLES specification tells REPORT which variables to include in the list.

Commands can begin anywhere on a line and continue for as many lines as necessary. That's why SPSS is so persnickety about that terminator (the period) — it's the only way it has of detecting the end of a command. The maximum length of a single line is 80 characters.

Keywords

All the commands in Syntax are keywords in the language. A *keyword* is a word already known to the language and has a predefined action. The variable names you define are not keywords, but SPSS can tell which is which by the way you use them. That is, you can name one of your variables the same name as one of the keywords, and SPSS can tell what you mean by how you use the word. Usually.

The names of commands, subcommands, and functions are keywords, and there are lots of them, but they are not reserved and you can use them freely. For example, you could have variables named format and report, and you could use the following Syntax command to display a list of their values:

```
REPORT /FORMAT=LIST /VARIABLES=REPORT FORMAT.
```

Don't try to name variables AND, OR, or NOT. These are logical operators in the Syntax language and, as such, are reserved words. If you try to use a reserved word as a variable name, SPSS will catch it and tell you that you can't do it. Relational operators are used in the Syntax language to compare values and are also reserved words. The relational operators are EQ, NE, LT, GT, LE, and GE. ALL, BY, TO, and WITH are also reserved words.

Variables and Constants

Most of the values used in Syntax are from the variables in the data set you currently have loaded and displayed in SPSS. You simply use one of your variable names in your program, and SPSS knows where to go and get the values for it. Some variables are already defined, and you can use them anywhere in your program. Predefined variables, which are called *system variables,* all begin with a dollar sign ($) and already contain values. The system variables are listed in Table 15-1.

Table 15-1	System Variables
Variable Name	**Description**
$CASENUM	The current case number. It's the count of cases from the beginning case to the current one.
$DATE	The current date in international date format.
$JDATE	The count of the number of days since October 14, 1582 (the first day of the Gregorian calendar).
$LENGTH	The current page length.
$SYSMIS	The system missing value. This prints as a period or whatever is defined as the decimal point.
$TIME	The number of seconds since midnight October 14, 1582 (the first day of the Gregorian calendar).
$WIDTH	The current page width.

You can create variables of your own to use as work areas to hold values while your program is running. These are called *scratch variables*. To create a scratch variable, use the # character at the beginning of the name. For example, the following command displays the number 34:

```
COMPUTE #FRED = 34.
PRINT / #FRED.
EXECUTE.
```

The PRINT command executes one time for each case (row) in the currently loaded data set, so it prints a line for each case. For example, if the data set contains 87 cases, the number 34 would be printed 87 times. If you were to include a variable name with the PRINT statement, all values of the variable would be printed. An EXECUTE statement is necessary following some commands — it's explained in detail later.

Data Declaration

You can define variables and their values in your program. To do so, you create a DATA LIST, which defines the variable names, and follow it with the list of values between BEGIN DATA and END DATA commands. In the following example, I created three variables and filled them with four instances of data:

```
DATA LIST / ID 1-3 SEX 5 (A) AGE 7-8.
BEGIN DATA.
001 m 28
002 f 29
003 f 41
004 m 32
END DATA.
PRINT / ID SEX AGE.
EXECUTE.
```

The DATA LIST command defines the variables. The first variable is ID. Its values are found in the input stream in columns 1 through 3, therefore it's defined as three digits long. It has no type definition so it defaults to numeric. The second variable is named SEX. It is one character long, and its values are in column 5 of the input. Its type is declared as alpha (A), so it's declared as a one-character string. The third variable, AGE, is two digits long, is a numeric value, and has its values in columns 7 and 8 of the input.

The BEGIN DATA command comes immediately after the DATA LIST command and marks the beginning of the lines of data — each line is a case. If you've ever wondered what it was like to place data on punched cards, this is it. SPSS is that old. This form of data entry still works, but this is the old way of getting data into SPSS. When this list of commands is executed, the normal SPSS window appears, containing the variable names and values. You can do all your processing this way, if you prefer.

But you don't have to do it by column numbers. You can enter the data in a comma-separated list, as follows:

```
DATA LIST LIST (',') / ID SEX AGE.
BEGIN DATA.
1,1,28
2,2,29
3,2,41
4,1,32
END DATA.
PRINT / ID SEX AGE.
EXECUTE.
```

END DATA must begin in the first column of a command line. It's the only command in Syntax that has this requirement.

Comments

You can insert descriptive text, called a *comment*, into your program. This text doesn't do anything except help make things clear when you read (or somebody else reads) your code. You start a comment the same way you start any other command: on its own line by using the keyword COMMENT or an asterisk. The comment is terminated by a period. For example:

```
COMMENT This is a comment and will not be executed.
```

An asterisk can be used with the same result:

```
* This is a comment placed here for the purpose of
describing what is going on, and it continues until
it is terminated.
```

You can also put comments on the same line as a command by surrounding them with /* and */. A comment like this can be inserted anywhere inside the command that a blank would normally go. For example, you could put a comment at the end of a command line:

```
REPORT /FORMAT=LIST /VARIABLES=SALARY /* The comment */.
```

It is important to note that the command is terminated with a period, but the period comes after the comment because the comment is part of the statement.

The Execution of Commands

Commands are executed one at a time starting from the top of the program. The order is important. In particular, if a variable has not been created yet, you can't use it. For the most part, the order is intuitive and you don't have to think much about what exists and what doesn't.

Some statements don't execute right away. Instead, they are stored for later execution. This is normally of no consequence because the statements will be executed when their result is needed. But you should be aware of it because it can cause surprises in some circumstances. For example, the PRINT command has a delayed execution:

```
PRINT / ALL.
```

This is a command to print the complete list of values for every case in your data set. It can print all the values, or by naming variables, it can print values of only the ones you choose. However, the PRINT command doesn't do it right away. It stores the instruction for later. When your program comes to a command that executes immediately, the stored commands are executed first. That works fine as long as there is a next statement, but if the PRINT statement is the last one in your program, nothing happens. That is, until you run another program, and the stored statement becomes the first one executed.

But there is an easy fix. All you need to do is end your program this way:

```
PRINT / ALL.
EXECUTE.
```

All the EXECUTE command does is execute any statements that have been stored for future execution. For the PRINT command there is another option. The LIST command does the same thing the PRINT command does, but it executes immediately instead of waiting until the next command:

```
LIST / ALL.
```

This execution delay may seem odd at first, but there's a reason for it. Many commands execute once for each case in your data. For example, if you have a series of three statements and you'd like a combination of the three executed once for each case, you need only enter the commands in your program in series. The commands will be stored and then executed, as a group, once for each case.

Flow Control and Conditional Execution

Unless you specify otherwise, a program starts at the top and executes one statement at a time through your program until it reaches the bottom, where it stops. But you can change that. Situations come up where you need to execute a few statements repeatedly, or maybe you want to skip one or more statements. In either case you want program execution to jump from one place to another under your control.

IF

You use the IF command when you have a single statement you want to execute only if conditions are right. For example:

```
IF (AGE > 20) GROUP=2.
```

This statement asks the simple question of whether AGE is greater than 20. If so, the value of GROUP is set to 2. We could have used the GT keyword in place of the > symbol. Table 15-2 lists the relational operators you can use to compare numbers.

Table 15-2		Relational Operators
Symbol	**Alpha**	**Definition**
=	EQ	Is equal to
<	LT	Is less than
>	GT	Is greater than
<>	NE	Is not equal to
<=	LE	Is less than or equal to
>=	GE	Is greater than or equal to

You can also combine the relational expressions with logical operators to ask longer and more complex questions. For example:

```
IF (AGE > 20 AND SEX = 1) GROUP=2.
```

This statement asks whether AGE is greater than 20 and SEX is equal to 1. If so, GROUP is set to 2. The logical operators are listed in Table 15-3.

Table 15-3		Logical Operators
Symbol	**Alpha**	**Definition**
&	AND	Both relational operators must be true
|	OR	Either relational operator can be true
~	NOT	Reverses the result of a relational operator

You should use parentheses to organize expressions so there is no ambiguity about what is being compared. When constructing a complicated conditional expression, it's easy to lose track of your original line of scrimmage.

You have to write your expressions so the computer knows what you're talking about. Spell them out. For example, IF (A LT B OR GT 5) is not valid. It can be written IF ((A LT B) OR (A GT 5)), which is a longer form but has a clearer meaning.

You can compare strings to strings and numbers to numbers, but you can't compare strings to numbers.

DO IF

The DO IF statement works the same way as the IF statement, but with DO IF you can execute several statements instead of just one. Because you can enter several statements before the terminating END IF, the END IF is required to tell SPSS when the DO IF is over. Following is an example with three statements:

```
DO IF (AGE < 5).
COMPUTE YOUNG = 1.
COMPUTE SCHOOL = 0.
END IF.
```

In addition to the ability to include a number of statements at once, you can use DO IF to test several conditions in a series and execute only the statements of the one that is true by using ELSE IF:

```
DO IF (AGE < 5).
COMPUTE YOUNG = 1.
ELSE IF (AGE < 9).
COMPUTE YOUNG = 2.
ELSE IF (AGE < 12).
COMPUTE YOUNG = 3.
END IF.
```

SELECT IF

The SELECT IF statement is not really flow control, but it works the same way. You can use it to remove cases and include only the cases you want in your analysis. For example, the following sequence of commands prints only the salary values greater than 40,000:

```
SELECT IF (SALARY > 40000).
PRINT / SALARY.
EXECUTE.
```

Any of the logical operators and relational operators that can be used in other IF statements can be used in SELECT IF statements.

DO REPEAT

If you want to perform a transformation on every value of a variable in a data set, the easiest way is to use DO REPEAT. For example, to increase the salary in every case by 10 percent:

```
DO REPEAT S=SALARY.
COMPUTE S=S + (S * 0.1).
END REPEAT.
PRINT / SALARY.
EXECUTE.
```

On the DO REPEAT command, the name of S is assigned as a stand-in for the values of the SALARY variable. The commands between DO REPEAT and END REPEAT are executed once for each value of SALARY — that is, once per case. Because S is the stand-in for each value of SALARY, any change you make to S is a change to one of the values of SALARY. At the end of this loop, every value of SALARY is printed.

Several lines with commands can be included between DO REPEAT and END REPEAT. Also, you can use several types of commands inside the loop, including IF, DO IF, and LOOP.

LOOP

With the LOOP command, you execute the same block of one or more statements repeatedly for a counted number of times. Following is a simple loop:

```
LOOP #LC = 1 TO 5.
COMPUTE #COUNT = #COUNT + 1.
END LOOP.
PRINT / #COUNT.
EXECUTE.
```

This program doesn't behave as you might first think, but it does give you an insight into the way the Syntax language works.

The first statement is a LOOP command and the scratch variable #LC is defined as a loop counter that runs from 1 to 5. The content of the loop defines another scratch variable, #COUNT, and adds 1 to it. Whenever a new scratch variable is defined, its original value is 0. Each time through the loop, 1 is added to #COUNT, so the value at the end of the loop — the value displayed by the PRINT statement — is 5.

But it doesn't stop there. This entire program is executed once for each case in the data set, so it's executed again and again. The value of #LC is always reset to 1, so the number of times through the loop is always 5. The second time through the loop the scratch variable #COUNT is already set to 5, so another 5 is added to it, resulting in 10 for the second line printed. The next line is 15, then 20, 25, 30, and so on for as many cases as you have in your data.

You can write the same program with the loop counter defined separately. The built-in loop counter is named MXLOOPS (short for maximum loops):

```
SET MXLOOPS = 5.
LOOP.
COMPUTE #COUNT = #COUNT + 1.
END LOOP.
PRINT / #COUNT.
EXECUTE.
```

But there's a problem doing it this way. You get warning messages in your output. The purpose of MXLOOPS is as a safety measure to prevent runaway loops, so it's best to specify the count in the LOOP command. Also, either of the following methods works for defining loop termination:

```
LOOP IF (#COUNT < 5).
COMPUTE #COUNT = #COUNT + 1.
END LOOP.
```

```
LOOP.
COMPUTE #COUNT = #COUNT + 1.
END LOOP IF (#COUNT > 5).
```

BREAK

You can use the BREAK command to stop a loop. For example:

```
LOOP #LC = 1 TO 5.
   COMPUTE #COUNT = #COUNT + 1.
   DO IF (#COUNT GE 12).
     COMPUTE #COUNT = 0.
     BREAK.
   END IF.
END LOOP.
PRINT / #COUNT.
EXECUTE.
```

In this example, every time the value of #COUNTER reaches 12 (or greater), the value is set back to 0 and looping stops. This program outputs 5, then 10, then 0, then 5, and so on, with one output line for every case in the data set.

Files

You can write data to files and read data from files. The simplest way to read files is to read SPSS-formatted files using GET.

GET

Whenever you choose File⇨Open⇨Data, SPSS issues a GET command to open an SPSS-formatted file and load it into SPSS. If you've loaded a file using the menu this way, you will have noticed in SPSS Viewer the GET command that loads the file. For example, the following program opens and loads the file named Cars.sav, and then changes the name of the data set:

```
GET
   FILE='C:\Program Files\SPSS\Cars.sav'.
DATASET NAME DataSet2 WINDOW=FRONT.
```

This command loads the data from the file, names it DataSet2, and opens a new SPSS main window displaying the data from the file in front of all the other windows. You don't need to ever load a file with the menu — you can load any file from within a Syntax program by specifying its name as the first argument to a GET command.

The quotes around the file name are optional, unless a blank is embedded in the name.

You don't have to load the entire contents of the file. If you want to omit certain variables, you can name them as part of the command, like the following:

```
GET FILE="Cars.sav" /DROP=MPG DISPLACEMENT.
```

You can even change the names of some variables. For example, the following changes MPG to MILESPERGALLON:

```
GET FILE='Cars.sav' /RENAME=MPG=MILESPERGALLON.
```

IMPORT

Files saved in the SPSS portable format can be copied from one type of computer to another and loaded into SPSS using the IMPORT statement. This type of file is in a format that is portable across all computers on which SPSS runs. To read such a file into SPSS, you use the following Syntax command:

```
IMPORT FILE=DATAFILE.
```

Any files created by EXPORT (or Save As in the portable format) from SPSS on any computer can be loaded by IMPORT into SPSS on any other computer.

SAVE

The SAVE command has the same result as choosing File⇨Save As and entering a file name. It writes the data to a file in the standard SPSS format. An example of the command follows:

```
SAVE OUTFILE='C:\Program Files\SPSS\Cars.sav'.
```

You have some options. You can specify DROP and RENAME the same as you can with the GET command. You can also compress the output file with the following option:

```
/COMPRESSED
```

EXPORT

The EXPORT command produces a portable data file containing the variables and data of the current data set. A file can be written with a statement like the following:

```
EXPORT OUTFILE=DATAFILE
```

Any files created by EXPORT (or Save As in the portable format) from SPSS on any computer can be loaded by IMPORT into SPSS on any other computer.

Chapter 16

Command Syntax Language Examples

Most Syntax command programs are short. That's because one command can do so much. This chapter is about the mechanics of writing and running programs. If you plan on doing much processing with SPSS, you'll certainly be doing some things over and over. If you save the procedures in a Syntax command program, you can just run the program instead of stepping through the process again.

Writing a Syntax Command Program

To write a new Syntax program, choose File⇨New⇨Syntax. The SPSS Syntax Editor dialog box appears, as shown in Figure 16-1, with a large blank text area. To write a program, type it into the blank area of the dialog box. To execute a program after you write it, choose Run⇨All.

Syntax programs are tightly tied to the variable definitions in the current data set because a Syntax program uses the data set's variable names, often in such a way that the type of the variable can be important. This means that the first instruction in a Syntax program is usually to load the data file.

You can load a file by choosing File⇨Open⇨Data or by writing a Syntax Command with a GET statement:

```
GET FILE='C:\Program Files\SPSS\Employee data.sav'.
```

Figure 16-1:
This is
where you
write Syntax
programs.

If you forget the exact form of this command, you can load a file using the menu and see the resulting command in SPSS Viewer. In fact, running any command by using the menu system causes its Syntax Command sequence to be written to SPSS Viewer.

The following is a program with a simple GRAPH command using the salary and job category information of the loaded data:

```
GRAPH TITLE = "Means of Salaries"
   /SUBTITLE = "separated by job category"
   /BAR = MEAN(salary) BY jobcat.
```

The resulting display in SPSS Viewer is shown in Figure 16-2.

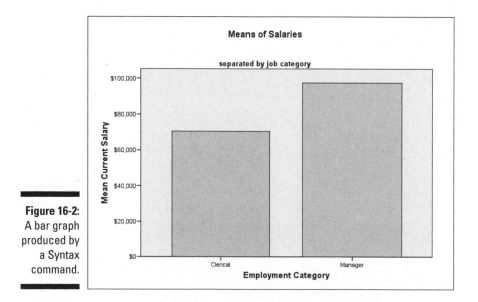

Figure 16-2:
A bar graph
produced by
a Syntax
command.

Saving and Restoring Programs

To load a Syntax program from disk, choose File⇨Load⇨Syntax from the menu of the main SPSS window. You will need to browse to the directory holding the file you want to load. Select the name and click the Open button; a new Syntax Editor dialog box appears containing the text of the program.

The ability to save a copy of your program is important. Whenever you write a Syntax program and think you may want to use it more than once, save it to disk so you can read it into SPSS and run it any time you want.

To save your program, you need to decide where you want to save it and what you want to call it. In the Syntax Editor dialog box, choose File⇨Save As and choose the location and name for the new file. If you've already saved the program (or if you loaded an existing program from disk), you need only choose File⇨Save to replace the existing file.

Often, you'll want to save your original program and create a new one by making changes to the original. In that case, load the original program from disk and then choose File⇨Save As to create a new file that holds your modified version. The original remains intact.

Adding a Syntax Program to the Menu

Every SPSS menu selection is nothing more than a command to execute a Syntax Command program. Adding a new item to the menu is a matter of adding a new menu button and assigning a task to it.

You can add new menu selections to customize SPSS and make it easier to do your common tasks. For example, if you are working on a data file and loading it regularly, you could define a new menu button to load the file for you. If you have an analysis or a report generation you run regularly, you could define a menu button that runs it with your set of parameters. Or you could set up a button to export data in your preferred format.

A menu consists of the menu bar (the part that's always visible at the top of the window), which contains a row of pull-down lists. Each list is made up of clickable buttons. Each button can be set to execute a Syntax command or to display another list of buttons. You can modify a menu by adding a new pull-down list or by adding a single button to an existing list. You can delete

existing menu items, but there is not any real need to do that — your modifications will almost always be for the purpose of adding buttons that perform
your own tasks.

The following steps take you through the process of adding a new pull-down
list with one button. The button executes the Syntax program named `load-
file.sps`, which is a program consisting of one GET statement to load a file:

1. **Create `loadfile.sps`.**

   Write and save a Syntax program that loads an SPSS data file. This
   example uses the one-line program, described earlier in this chapter,
   that uses the GET command to load `Employee data.sav`.

2. **Choose Utilities⇨Menu Editor.**

   The dialog box shown in Figure 16-3 appears. This menu choice can be
   made from any of the system menus — the main SPSS dialog window,
   SPSS Viewer, or even Syntax Editor. Any of the menus can be modified
   through this dialog box.

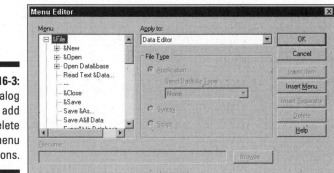

Figure 16-3:
The dialog
box to add
and delete
menu
selections.

3. **In the Apply To pull-down list, select Data Editor.**

   This is the choice of which menu to modify. The other three choices
   are View, Script, and Syntax. Each time you choose a different menu,
   the buttons that are already defined for that menu show up in the Menu
   box on the left. Initially all four menus are identical, so the Menu text
   doesn't change.

4. **In the list of names in the Menu box, click the plus sign next to
   &Open.**

The list expands to display the items already defined for the Data Editor menu: D&ata, &Syntax, &Output, S&cript, and End Of &Open Menu. The ampersand (&) in the name specifies the following letter as the shortcut key that activates the menu item. You can include an ampersand in the name you add, if you want.

If you use an ampersand to specify the same letter as a shortcut for more than one menu selection, SPSS will use one and ignore the other — which is probably not what you intended.

5. **Select End Of &Open Menu.**

The selection becomes highlighted. Whenever an item is added to the menu, it is added immediately *before* the selected item. (The End entry is included in the list only so the last position can be selected.)

6. **Click the Insert Item button.**

A new menu button appears with the name New Menu Item.

7. **Type the name** MyFile **and press the Enter key.**

The text you type replaces the name of the selected menu item.

8. **In the File Type area, select Syntax.**

The new menu selection can be associated with another application or a script, but in this example the new menu selection will be executing a Syntax program.

9. **Click the Browse button and locate the Syntax program file.**

Where to find Syntax commands

There are lots of Syntax commands, and they all have lots of options. If you have something you want to do, and you want to find the Syntax command to do it, you have two basic approaches.

One way is to use the menu system to command SPSS to do whatever it is you would like it to do. In SPSS Viewer, you will be able to see the text of the Syntax commands that generated the output. Highlight that text, choose Edit⟶Copy, switch to the Syntax Editor dialog box, and choose Edit⟶Paste to capture the text into your own program.

Another way to find commands is to use the help menu in SPSS. It may take a few tries to get the Syntax command you want, but it's listed in there somewhere. The Syntax commands are listed in all uppercase letters, so they're easy to spot.

Clicking the button opens a browse window. Locate the directory containing the file `loadfile.sps`. To make the file name appear, you may need to choose Syntax Files (*.sps) in the Files Of Type pull-down list at the bottom of the dialog box.

10. Click the Open button.

This attaches the file to your new menu item.

11. Click the OK button.

The Open menu of the main window of SPSS has been modified by the addition of MyFile, as shown in Figure 16-4.

If you want to add the same option to the menus of other dialog boxes in SPSS, you have to follow the same procedure for each one. Adding a menu item takes only a small amount of work and can prevent many repeated steps. For example, if you're in the process of entering and correcting data, a simple menu item to load the file would keep you from hunting for it every time you need to load it. Also, if you have a group of analyses you run repeatedly, you could include them all in one Syntax program and have them all run for you at the click of a button. The same program could load the file at its start, so you only need to click one button to do all your work.

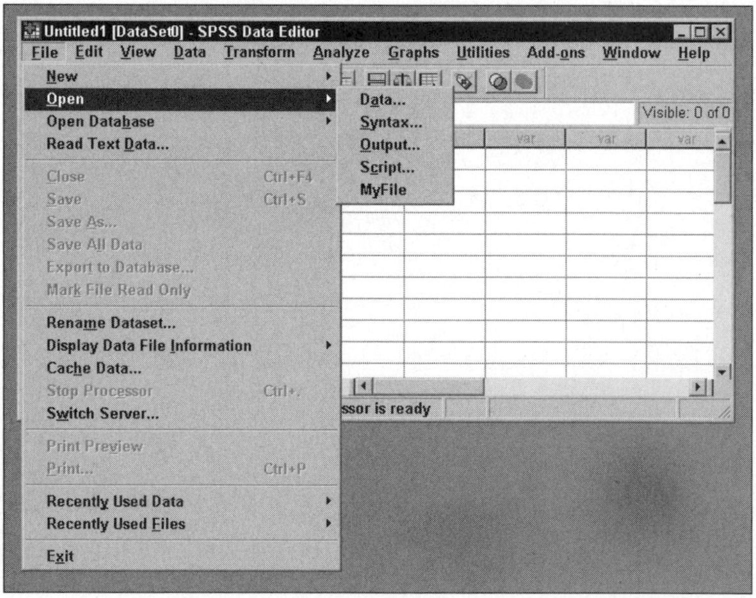

Figure 16-4:
The MyFile
selection
has been
added to the
Open menu.

Doing Several Things at Once

You can write a Syntax program to do more than one thing. All the commands in one script are executed one after the other. And because one Syntax command can do quite a bit, you don't have to write much of a program to do lots of processing. For example, the following four-line program named `makeplot. sps` performs four separate tasks:

```
GET FILE='C:\Program Files\SPSS\Cars.sav'.
DATASET NAME DataSet1 WINDOW=FRONT.
GRAPH LINE=MEAN (HORSE) BY YEAR.
GRAPH BAR=MEAN (MPG) BY ACCEL.
```

The first line loads the SPSS data file named `Cars.sav`. The second line renames the data set to `DataSet1` and brings the window displaying it to the front. As a result, if data has already been loaded and named `DataSet`, this new file will assume the name (the other will be closed). The last two lines draw graphs — one line graph and one bar graph, as shown in Figures 16-5 and 16-6. Note the way in which variable references are made on the `GRAPH` commands. Referring to a variable by its name results in all its values being used; using the word `MEAN` before the variable name in parentheses results in the mean of the variable's values being used. These commands are simple but the actions are complex.

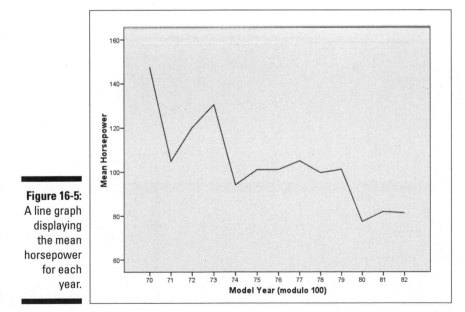

Figure 16-5:
A line graph displaying the mean horsepower for each year.

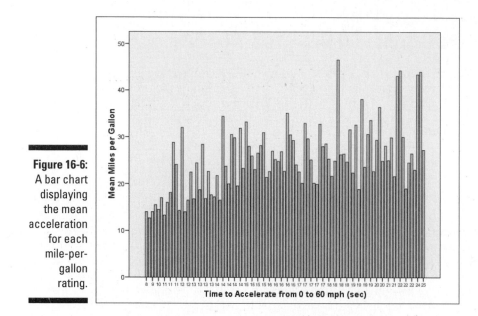

Figure 16-6:
A bar chart
displaying
the mean
acceleration
for each
mile-per-
gallon
rating.

Graphing P-P and Q-Q Plots

The Syntax language contains the PPLOT command, which can be used to generate either a q-q plot or a p-p plot. The following program, named makeplot2. sps, contains commands to produce both.

```
GET FILE='C:\Program Files\SPSS\Employee data.sav'.
DATASET NAME DataSet1 WINDOW=FRONT.
PPLOT SALARY
  /TYPE=Q-Q.
PPLOT SALARY
  /TYPE=P-P.
```

This program loads the data set and then produces a plot of each type. The q-q plot is displayed in Figure 16-7. A q-q plot is a quantile-quantile plot, in which the quantiles of the actual values are plotted against the quantiles of the expected values.

Figure 16-8 displays the p-p plot produced from the program. A p-p plot is a proportion-proportion plot, in which the actual proportions are plotted against the expected proportions.

Figures 16-7 and 16-8 do not represent all the output you get from the PPLOT command. In particular, a detrended plot (a plot in which the actual values are plotted against deviations of the expected values) is also produced.

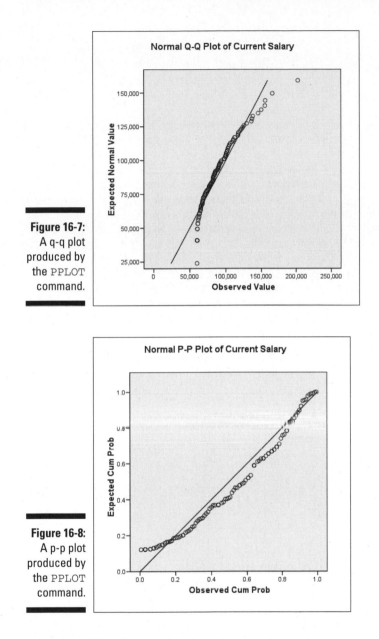

Figure 16-7:
A q-q plot
produced by
the PPLOT
command.

Figure 16-8:
A p-p plot
produced by
the PPLOT
command.

Splitting Cases

In this section we look at a program that loads a data file and counts the
repetition of values in a certain variable. The repetition count is made for all
cases in a file, and then the file is split and a count is taken for each portion.
The program is named splitfile.sps:

```
GET FILE='C:\Program Files\SPSS\Employee data.sav'.
FREQUENCIES SALARY.
SORT CASES BY GENDER.
SPLIT FILE BY GENDER.
FREQUENCIES SALARY.
```

The first line of the program uses the GET command to load the file. The second line uses the FREQUENCIES command to generate the counts and percentages for the salary values. The top section of the table produced from this command is shown in Figure 16-9. As you can see, the table generated includes five columns. A salary value is shown in the first column and the count of the total number of occurrences of the value shown in the second column. The Percent column holds the percent of the total number of cases (excluding cases with missing values in any variable) that contain this particular salary value. The Valid Percent column holds the percent of the total number of cases (including those with missing values in other variables) that contain this particular value. The Cumulative Percent is the number of cases with salaries less than or equal to the salary shown in the first column. For this example, the values displayed as Percent and Valid Percent are the same because none of the cases in the displayed portion contain a missing value for any variable.

Current Salary

		Frequency	Percent	Valid Percent	Cumulative Percent
Valid	$60,075	1	1.0	1.0	1.0
	$60,300	4	3.8	3.8	4.8
	$60,525	2	1.9	1.9	6.7
	$61,200	2	1.9	1.9	8.7
	$61,650	1	1.0	1.0	9.6
	$62,325	1	1.0	1.0	10.6
	$63,000	1	1.0	1.0	11.5
	$63,450	2	1.9	1.9	13.5
	$64,500	1	1.0	1.0	14.4
	$65,115	1	1.0	1.0	15.4
	$65,250	1	1.0	1.0	16.3
	$65,475	1	1.0	1.0	17.3

Figure 16-9: A frequency table for the entire data set.

The SORT command is used in the program to sort cases. You must sort a data set on the variable about to be used to split a file because variables of like values must all be together for the split to work properly. In this example, the SORT command will group all the female cases before the male cases.

The SPLIT command logically inserts dividers at each point where the value of the named variable changes. In this example, the value f is used for female and the value m for male, so a logical divider is placed between them. The divider is logical because the split refers only to the memory-resident form of the data — the split does not survive the data being written to a file.

The last line of the program builds a new set of counts and percentages, but this time the data is divided by gender, so the table is generated in two parts. The upper part of the table is shown in Figure 16-10. The headings of the table have the same meanings they had before, but you can see that the top of the table contains the numbers for the female cases and the bottom portion contains data from males. If the SPLIT command had used a variable with more values, the cases would have been split into more parts.

Current Salary

Gender			Frequency	Percent	Valid Percent	Cumulative Percent
Female	Valid	$61,200	1	9.1	9.1	9.1
		$62,325	1	9.1	9.1	18.2
		$64,500	1	9.1	9.1	27.3
		$65,250	1	9.1	9.1	36.4
		$70,313	1	9.1	9.1	45.5
		$76,500	1	9.1	9.1	54.5
		$81,000	1	9.1	9.1	63.6
		$81,563	1	9.1	9.1	72.7
		$83,625	1	9.1	9.1	81.8
		$85,125	1	9.1	9.1	90.9
		$87,188	1	9.1	9.1	100.0
		Total	11	100.0	100.0	
Male	Valid	$60,075	1	1.1	1.1	1.1
		$60,300	4	4.3	4.3	5.4
		$60,525	2	2.2	2.2	7.5
		$61,200	1	1.1	1.1	8.6
		$61,650	1	1.1	1.1	9.7
		$63,000	1	1.1	1.1	10.8
		$63,450	2	2.2	2.2	12.9

Figure 16-10: Separate frequency tables for females and males.

Examining Data

The EXAMINE command in the Syntax language may be the quickest way to look at data. For example, with the system data file named Cars.sav loaded into SPSS, a two-word Syntax program produces a graph of a variable. The two-word program is as follows:

```
EXAMINE MPG.
```

This command results in the box plot shown in Figure 16-11, which graphically displays the mean, the standard deviation, and the extreme values.

But that's not the only way EXAMINE can show you data. You can include more than one variable, or you can change the plot style to a histogram. The following command generates more than one histogram:

```
EXAMINE ACCEL, HORSE /PLOT=HISTOGRAM.
```

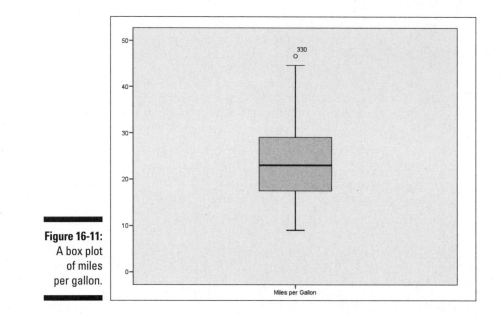

Figure 16-11:
A box plot
of miles
per gallon.

This command produces a histogram for each of the two named variables. The histogram representing the acceleration values (ACCEL) is shown in Figure 16-12.

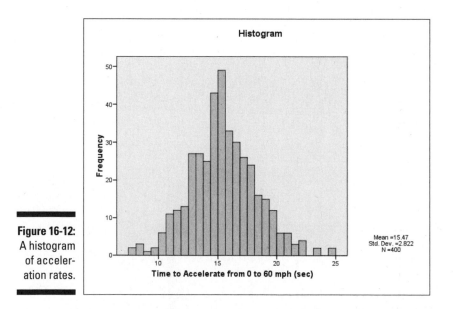

Figure 16-12:
A histogram
of acceler-
ation rates.

Part VI

Programming SPSS with Python and Scripts

The 5th Wave   By Rich Tennant

Well, heck — that's just darn impressive! And you say it's programmed to sew up and dress the incision afterward as well?

In this part . . .

You can write commands using the Python programming language and include them among the Command
Syntax statements. The result is the same as if you had
written Command Syntax, but Python is a more modern
and flexible language. If you think you might want to
become a computer nerd, this is the way to go.

And you can use scripting — programming in Sax BASIC —
to create programs that execute automatically.

Chapter 17

The Python Programming Language

ython is a general-purpose programming language, and it has been added to SPSS version 15 as a scripting language. Chapter 20 is about using Python inside SPSS, but this chapter is about the Python programming language itself. If you're not a programmer, don't worry about it. Python is famous for being easy to learn. And you might think it's named after a snake, but it isn't. It's named after *Monty Python's Flying Circus*. I just thought I'd mention that in case you thought things were going to get serious. Now for something completely different.

You Type It In and Python Does It

If you give Python an instruction that it understands, it will obey it and do something. It's very obliging that way. But you have to be specific when you tell it what you want it to do.

If you want a Python of your own, outside of the one that comes with SPSS, you can download and install one from the Internet for free. By playing with your own Python, you can see how the examples in this chapter work. The only way to really learn a programming language is to fiddle around with it and write some programs of your own. Sometimes you get great insight into programming from finding out what doesn't work.

Python is an *interpreter*. That is, instead of taking your set of program instructions and translating them to machine language, it just reads and obeys whatever you type. That is, it reads your commands like it would read a script, so you will also hear Python programs called *scripts*.

Python can be used to generate graphic displays, communicate over the Internet, make calls into the operating system, and other things that we won't be messing with. This chapter shows you just enough basic Python to get you comfortable writing scripts for SPSS.

When you fire up the stand-alone version of Python, it displays >>> as a prompt for you to give it some instructions. If you type something it knows how to do, it will do it. If you type something it doesn't understand, it will complain — but it won't bite. Remember, it's not a snake.

The Way Python Does Arithmetic

Statistics is made out of arithmetic, and Python is good at arithmetic. You can enter any expression you want, and Python will do the calculations and give you the answer.

Let's start with something simple. At the prompt, type a simple addition such as the following. Python comes back with the result:

```
>>> 2 + 2
4
```

You can use multiplication, division, decimal points, parentheses, and all sorts of fancy stuff:

```
>>> (88 + 2) / 6
15
```

The symbol for multiplication is the asterisk:

```
>>> 10 * 10
100
```

If you do integers, Python does integers. If you do decimal points, Python does decimal points. Integer arithmetic just chops things off like this:

```
>>> 7/2
3
```

And arithmetic using decimal points (floating-point arithmetic) keeps the fractional portion, like this:

```
>>> 7.0/2.0
3.5
```

You can mix integer and decimal numbers in the same expression, but you need to watch what you're doing. Whenever any operation involves at least one number with a decimal point, Python treats all the numbers as if they have decimal points. For example:

```
>>> 7/2.0
3.5
```

You need to be careful when you mix the number types like that. You could get something other than what you expect. The following two examples look similar, but they are actually different:

```
>>> 7.0/2.0 + 4.5
8.0
>>> 7/2 + 4.5
7.5
```

The first example performs a decimal point division and winds up adding 3.5 to 4.5. The second example performs an integer division, which chops off the decimal part, and winds up adding 3 to 4.5. These results are different in the sense that one is wrong for whatever you happen to be calculating.

Use decimal points in all your numbers unless you have a specific reason not to. That way, nothing gets chopped off and thrown away.

Instead of just printing the numbers on the display, as we've done so far, you can store them in a name, called a _variable_. The three dimensions of a box could be stored in variables this way:

```
>>> height=20.0
>>> width=9.0
>>> depth=12.0
```

No number is displayed this time. If you store a number somewhere, Python doesn't display it. Python remembers those names and numbers for you. You can calculate the volume of the box and have it displayed this way:

```
>>> height * width * depth
2160.0
```

The equal sign (=) is the assignment operator. It takes the value from whatever you put on the right and stores it in whatever location you name on the left. It simply writes over whatever was there before.

If you want, you could store the volume in another variable and then display it, like this:

```
>>> volume = height * width * depth
>>> volume
2160.0
```

Whatever name you enter is the one Python uses. If you spell it wrong, it's a different name, so use names that are easy to spell. And don't use things like the uppercase letter I and the lowercase letter l because they can be confused with one another and with the number 1. And watch out for the letter O and the number 0.

Python has the memory of an elephant snake. After you stick a value in a variable, it will remember it forever. Well, at least until you end the program. If you want to really save a value, you need to write it to a file on disk so you can read it back. That's easy to do, and we get to it later.

As you have seen, if you simply name a value or a variable, Python prints it for you. You can also use the `print` command, like this:

```
>>> print volume
2160.0
>>> print height,width,depth
20.0 9.0 12.0
```

As you can see, Python remembered. And you can see that the `print` command can handle more than one value at a time.

The Way Python Handles Words

If you want Python to notice what you're saying, you will need to put it in quotes. You can use either single quotes or double quotes, but whichever one you use at the start is the one you must use at the finish. Like this:

```
>>> 'Single quotes'
'Single quotes'
>>> "Double quotes"
'Double quotes'
```

If you enter a quoted string by itself this way, Python echoes it back to you just like it does a number. Python usually uses single quotes when it echoes, but that's just an attitude problem and doesn't matter.

In the world of computer programming, any group of characters used to make up a name or a sentence or anything you can read is called a *string*. Also, a blank is a character just like any other, except you can't see it if you're a mere mortal.

You can put single quotes inside double quotes and double quotes inside single quotes, like this:

```
>>> "Girl's clothes?"
"Girl's clothes?"
>>> '"Girl clothes?" he asked'
'"Girl clothes?" he asked'
```

Hmm. This time Python uses double quotes to display the string that contains a single quote. Attitude meets necessity. Don't think about it too much. Let's move on to an example of storing a string in a variable:

```
>>> fred="Is this a cheese shop? "
>>> fred
'Is this a cheese shop? '
```

You can stick a string in a variable exactly the way you can a number. You can even add one string to another one, like this:

```
>>> herbie = fred + "Is this a parrot shop?"
>>> herbie
'Is this a cheese shop? Is this a parrot shop?'
```

As you can imagine, the strings can get long. You can make them show up on more than one line by inserting a \n (newline) character and using the print command, like this:

```
>>> herbie = herbie + "\nNo. This is for lumberjacks."
>>> print herbie
Is this a cheese shop? Is this a parrot shop?
No. This is for lumberjacks.
```

The print command translates \n as being the start of a new line. If you just echo the variable, it doesn't work — you just get the two characters \n in the output.

Now for something slightly different. Using triple quotes causes the automatic insertion of newline characters into your string whenever you start a new line. You can organize formatted text with it, like this:

```
>>> hebert="""
... Algy met a bear
... The bear was bulgy
... The bulge was Algy
... """
>>> print hebert
Algy met a bear
The bear was bulgy
The bulge was Algy
>>>
```

Notice that Python drops the normal >>> prompt while you are entering the triple-quoted string and uses three dots (...) instead. It's not important — it's just another example of Python assuming an attitude.

You can use either single quotes or double quotes to construct your triple quotes. If that sentence makes any sense to you, you're really getting into this. Let's move on. I showed you earlier how you can add strings; now I'll show you how they can be multiplied:

```
>>> essword="spam "
>>> print essword * 7
spam spam spam spam spam spam spam
```

If you want to define a long string, you can break it and enter it on more than one line, like this:

```
>>> go="Now is the time for all good men to\
... get out of town."
>>> print go
Now is the time for all good men toget out of town.
```

When you are entering a string of characters, you can put a backslash (\) at the end of the line and continue at the beginning of the next line just as if you had continued on the same line. As you can see by toget in the output line, I should have added a space after to and before the backward slash. You can also build long strings by adding smaller strings without putting in a plus sign.

```
>>> hank="ugly "    'dog'
>>> hank
'ugly dog'
```

You might want to do it that way and you might not. I think a plus sign between the two makes it a lot clearer, but you might want to leave it out just to show off. That's what I was doing when I put this example in the book.

Okay. That's enough about putting strings together. Let's take some apart. It's easy because you can refer directly to each letter by its position number. The letter at the extreme left is number 0, the next one is number 1, and the next one is number 2, and so on. For example, to pull the first letter out of the string of the preceding example, you just address it by number, like this:

```
>>> hank[0]
'u'
```

If you want to extract a range of characters, just use the number of the first character you want and the number of the character *following* the last one you want, and put a colon in between the two, like this:

```
>>> hank[2:6]
'ly d'
```

If you use the colon but leave out the first number, Python assumes 0 and starts at the first character on the left. If the ending number is missing, it assumes the end of the string. For example:

```
>>>> hank[:4]
'ugly'
>>> hank[5:]
'dog'
```

You can use extraction to build new strings by adding the pieces together like this:

```
>>> frank = 'very ' + hank[:4] + ' fat ' + hank[5:]
>>> frank
'very ugly fat dog'
```

One of the questions that always comes up in a program is, "How long is that string?" Here's how to find out:

```
>>> len(hank)
8
>>> len(frank)
17
```

You will find lots of functions that do things to strings that result in new strings that are different. The original string is never changed — you can't change an existing string no matter what you do. To make a difference in a string, you have to create a new string and replace the original. Here are a bunch of examples of functions doing things to strings:

```
>>> hank.capitalize()
'Ugly dog'
>>> hank.find("dog")
5
>>> hank.replace('g','x')
'uxly dox'
>>> hank.title()
'Ugly Dog'
>>> hank.upper()
'UGLY DOG'
```

Remember, none of these examples changed the original. They produced new strings. But this group of functions is just the tip of the iceberg, or should I say stringberg? You will find a Python function to do just about anything you can imagine to a string. It even does some things that would, under normal circumstances, be considered too personal.

The Way Python Handles Lists

You can have a variable hold an arbitrary collection of strings and numbers. You address any specific one by its position number in the list, with the first one in the list being number 0, like the following:

```
>>> jam=['a',100,"c",'dee']
>>> jam
['a', 100, 'c', 'dee']
>>> jam[0]
'a'
>>> jam[1]
100
>>> jam[1:3]
[100, 'c']
```

In this example, you can see where four things were stuffed into the variable named jam. When the variable was displayed, all four items it contained were displayed. However, by using a position value to refer to individual items in the list, it is possible to address one item at a time. Or, as in the case of the last example, it is possible to select a subset of the items in the list.

When you use a pair of position numbers, the first number is the number of the first item you want, but the second number is the number *following* the last item you want. Also, the first item in the list is always number zero.

The position values on lists work like the position values on strings, but you may recall from earlier that strings can't be modified. Lists can. You can replace one member of a list by simply assigning a new thing to it, like this:

```
>>> jam = ['a', 100, 'c', 'dee']
>>> jam
['a', 100, 'c', 'dee']
>>> jam[2]='hooha'
>>> jam
['a', 100, 'hooha', 'dee']
```

You can quickly find out how many things are in a list:

```
>>> len(jam)
4
```

Lists are one of the really nice things about Python. If you want to do something to a list, try it. It will probably work. You can even put lists inside lists:

```
>>> jam[0] = ['apple', 'pear']
>>> jam
[['apple', 'pear'], 500, 'hooha', 'dee']
```

Making Functions

Python can remember a set of instructions for you, and you can later call on that set by name. Here's a simple example that divides a number in half and displays the results:

```
>>> def showhalf(x)
...      print x/2
```

The line with the `def` command names this as a function called `showhalf`. This example has one variable, named `x`, that is used in the body of the function. All statements following the definition line will be included as part of the function, as long as you indent them. After you type a line that is not indented by the same amount or more, the function ends. Python then remembers your definition of the function and you can use it as often as you like. For example:

```
>>> showhalf(10)
5
>>> bunch=100
>>> showhalf(bunch)
50
```

Whatever value you include in the parentheses becomes the value of x inside the function when it is called on to do its thing. The following shows that, instead of just doing something inside, as in the previous example, the function can return a value to you:

```
>>> def getthird(value):
...     return(value / 3.0)
...
>>> j = 9
>>> k = getthird(j)
>>> print k
3.0
```

In this example, whatever value is passed to the function is divided by 3 and the result comes back because it is part of a `return` statement. You can pass anything into a function and return anything else: strings, numbers, lists, whatever you want.

It's normal to have a Python program begin with a bunch of function definitions and then have the body of the program use the functions to do its work. Functions can even call other functions, but be careful. Too much disorganization leads to something called "spaghetti code," which can become so convoluted you can't read it.

You should know that although you can get only one value back from a function, you can pass lots of values to one. Here's an example of a function needing more than one value for its input:

```
>>> def showsum(a,b,c):
... print a+b+c
...
>>> showsum(3,5,9)
16
```

The limit of being able to return only one value from a function is never a problem. If you find that you need to return more than one value, you can just return a list, but in reality you probably need more than one function.

Here's a nifty trick. You can define your function to have some defaults for some of the values you pass to it. Then, if you leave out any of those values when you invoke the function, the defaults will be used:

```
>>> def spark(a,b="too big",c=44)
...     if a > c
...         print b
...
>>> spark(20)
>>> spark(50)
too big
>>> spark(100,"way too large")
way too large
```

In this example, the function named `spark()` has three arguments; a, b, and c. The last two have default values. The function simply tests whether the value of a is larger than the value of c, and if so it prints b. In the example, the first call to the function sets the value of a to 20, which is not larger than c, so nothing happens. In the second call, a is set to 50, and that's larger than c, so the default content of b is printed. The last call to `spark()` has a value for a that is larger than c, but the string printed for b is different because the value passed to the function overrides the default.

Function definitions are, in a way, the heart of the system. You normally write a program by defining your own functions and using them along with Python's plentiful built-in functions. This program structure becomes particularly convenient when you do the same sort of thing more than once, but the most important characteristic of this program structure is that it makes it possible to organize your instructions in a logical way. The main problem with programs is not writing them — it's fixing them later when they don't work the way you want. And the main problem with fixing them is finding out where to make the change. Be organized!

Asking Questions with `if`

Often, you'll have a statement or two that you want to execute only under certain conditions. You can use an `if` statement to ask the question, and the indented statements following it will be executed only if the answer to your question is true. For example:

```
>>> x = 3
>>> if x < 5:
...     x = 20
...     print x
...
20
```

You group statements together and have them all execute as a single unit by putting them together as a *block*. A block is created when two or more consecutive statements are indented by the same amount.

Sometimes you'll want to do one thing under some circumstances and something different under other circumstances. That's where you can use `else`:

```
>>> x = 10
>>> if x < 8:
...     print 'x is less than 8'
... else:
...     print 'x is not less than 8'
...
x is not less than 8
```

In this code, instead of just ending the statements in the `if` block, the `else` keyword, followed by a colon, is used to start a new block. When the code was executed, the first block was skipped but the second was run. Using `if`/`else` statements this way means that one, and only one, of the two blocks of code execute.

Using `if` blocks in code is common. If you write a script of any complexity, you will nest such blocks inside one another. With a bit of practice, you will get proficient at doing such things. However, one odd situation comes up, usually when you back up to change something. You will find yourself needing to put in some code that does nothing at all. Python is persnickety about its syntax, and there are places where you are always required to put in something, but you may find that you don't want to do anything. To the rescue comes the keyword `pass`, which you can use like the following:

```
>>> x = 3
>>> if x < 8:
...     pass
... else:
...     print 'x is not less than 8'
...
```

This example has no output because all it does is execute the `pass` command, which does absolutely nothing. But perhaps you want to use the `if` statement to select a single action among several possible choices. You can do that as follows:

```
>>> x = 8
>>> if x < 8:
...     print 'x is less than eight'
... elif x == 8:
...     print 'x is equal to eight'
... else:
...     print 'x is greater than eight'
...
x is equal to eight
```

The `elif` keyword is short for *else if* and allows you to add another condition that must be true followed by another block of statements that will be executed only if that second expression is true. You can daisy chain as many of these `elif` statements as you want, and only the first one found to be true is executed — the rest are skipped. You can have only one `else` statement, and it must come last.

Be sure you say what you mean. While the single equal sign (=) is the assignment operator and is used to copy data, the test for a couple of values being equal is the double equal sign (==). You can include the *greater than or equal to* test with >=, the *less than or equal to* test with <=, and the *not equal to* test with !=.

You can also use `and`, `or`, and parentheses in expressions. Things can get complicated if you need them to. For example, the following is true only if `aa` is greater than or equal to `bb` *and* `x` is not equal to `y`:

```
>>> if (aa >= bb) and (x != y):
```

Don't think too much about what that statement means. It just leads to headaches. I wanted to show it to you so you'd know that that sort of thing is possible if you really need it, or if you find yourself with a sudden urge to do something baroque.

Doing It Over Again with for *and* while

One of the things often done in programming is repetition. Having your program go back through the same code again is called *looping*, or *iteration*. (You're probably familiar with the word *reiterate*, which means to repeat something.)

You can iterate in Python by using the `for` keyword, like this:

```
>>> bog = ['first',50,'third',800,3.14159]
>>> for x in bog:
...     print x
...
first
50
third
800
3.14159
```

You first create a list and then set up a variable in the `for` loop to iterate through the list. The loop executes once for each member of the list, with the variable assuming, for each iteration, the value of a member of the list. It couldn't be easier. Well, if you think of an easier way, tell the folks at Python and I'm sure they'll put it in the language.

Don't change any of the values in the list while you're inside the loop. The results are unpredictable, and the last thing you want in your computer is a confused Python. If you absolutely positively have to change the list inside the loop, use a copy of the list to iterate.

It is more common in other programming languages to iterate a specific number of times. You can do that in Python if you feel you must. A special built-in function called `range()` returns a list and lets you iterate a set number of times. You can do it this way if you feel an irresistible urge to count:

```
>>> for z in range(5):
...     print z
0
1
2
3
4
5
```

Or you can use the `range()` function for starting at some value other than 0, like this:

```
>>> for y in range(5,10):
...     print y
5
6
7
8
9
```

Iterating by a count is actually not a different capability of the language — the `range()` function simply returns a list containing the numbers needed for the count. But a different capability of the language *is* found in the other iterater, named `while`. It works a lot like `if`, except it repeats continuously, testing a conditional expression to determine when to stop. `while` continues to execute its block of statements as long as the condition it tests comes up true. The following is a simple example:

```
>>> x = 2
>>> while x < 8:
...     print x
...     x = x + 1
...
2
3
4
5
6
7
```

When inside a `while` loop, make sure you do something that affects the value of the expression tested by the `while` command. Otherwise, you could be caught in the loop forever. And that's an embarrassingly long time.

I said earlier that a while statement is sort of like an if statement. In fact, it is so much like an if statement that you can put an else at the end of the block of a while statement, like this:

```
>>> x = 7
>>> while x < 9:
...     print x
...     x = x + 1
... else:
...     print "The loop is done"
...
7
9
The loop is done
```

The first part of the loop works just like an if statement, except it executes over and over as long as the conditional expression is true. Once the expression becomes false, the else part of the statement executes once and then the while statement is finished.

"But, hold varlet," you shout, drawing your sword. "A statement following the loop would execute once without regard to the presence of else." Whereupon I wisely retort, "Stay your hand. Bear with my discourse but a bit longer and I will show you purpose." Then I cleverly explain the operations of continue and break.

A continue statement anywhere inside a for loop or a while loop will cause the rest of the statements inside the loop to be skipped. That is, the continue keyword jumps immediately to the bottom of the loop, allowing things to come back around again normally.

A break statement inside a loop will cause the while or for loop to be abandoned as if all iterations had completed, whether or not that is the case. In fact, when a break statement abandons the execution of a loop, it will also cause any terminating else code *to be skipped*. This is where you slip your sword back into its scabbard while muttering, "I'll get you next time."

"One more thing!" I shout. "It is common to nest for and while loops inside one another. When that happens, the continue and break statements only continue or break the innermost loop." I mention this only because it's the kind of thing that can send you on a long fruitless bug hunt.

Chapter 18

Python inside SPSS

· ·

In This Chapter

▶ Installing Python

▶ Running a Python program inside SPSS

▶ Running Python combined with Syntax

· ·

*T*his chapter is the gateway to becoming an SPSS power user. It contains the mechanics you need to know to write Python programs that run inside SPSS. Python is a programming language apart from SPSS, and integrating Python with SPSS makes it possible to do some things that would otherwise be difficult to do in the Syntax language. To use Python, you need to know the basics of the SPSS Syntax command language because you actually reach out of Python into the Syntax language to issue commands to SPSS. You can think of the Python plug-in as an extension of the built-in Syntax language.

Python was designed to be a general-purpose language, so it has a much larger scope than you will ever need for SPSS programming. This large scope means that it contains features and capabilities you will never use within SPSS. On the other hand, it also means that you can solve special problems unique to your situation.

Don't let Python's size intimidate you. It's sort of like having a pocket calculator with lots of extras — if you see a button that doesn't make sense to you, ignore it.

Installing Python for SPSS

Python is not installed as part of the SPSS base system. You have to install it separately. It's on the SPSS CD and can be installed with the following steps:

1. Insert the SPSS CD.

The main installation dialog box is displayed.

2. **Click the Install Python 2.4.3 selection.**

   After a short pause (during which nothing seems to happen), the license agreement appears, as shown in Figure 18-1.

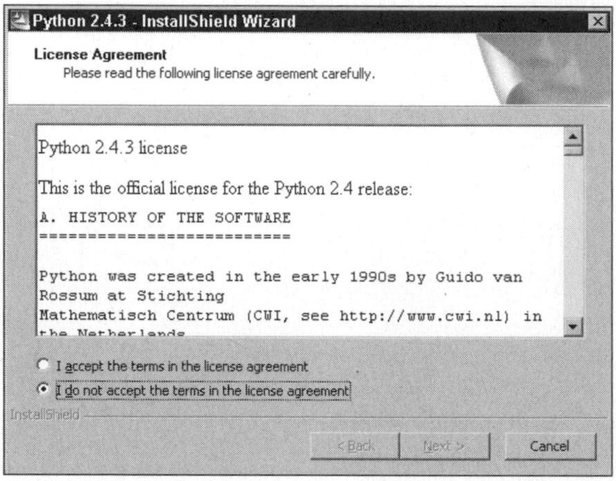

Python 2.4.3 - InstallShield Wizard

License Agreement
Please read the following license agreement carefully.

Python 2.4.3 license

This is the official license for the Python 2.4 release:

A. HISTORY OF THE SOFTWARE
==============================

Python was created in the early 1990s by Guido van Rossum at Stichting Mathematisch Centrum (CWI, see http://www.cwi.nl) in the Netherlands.

○ I accept the terms in the license agreement
● I do not accept the terms in the license agreement

InstallShield

< Back   Next >   Cancel

Figure 18-1:
The Python 2.4.3 license agreement.

3. **If you decide to continue after reading the license, click the I Accept the Terms... option and then click the Next button.**

   You should read the license agreement before you accept it because you will be bound by it. A progress bar dialog box appears as files are copied to your disk.

4. **Choose whether you want to allow Python to be accessible to other logins on the computer (see Figure 18-2) and then click Next.**

   Install Python for all users unless you have a specific reason to exclude someone.

5. **Select the name of the directory to contain the Python files, as shown in Figure 18-3, and then click Next.**

   Although you can choose any directory name and disk on your system, I suggest you use the default directory, Python24. If you do use a different directory, don't use an existing one with other files already in it.

6. **Decide whether or not to leave out portions of the Python files.**

   Unless you have a specific disk space problem, accept the default, as shown in Figure 18-4, and install it all. If you have a disk space problem, click Disk Usage to see how much space you have (and which disks have enough space). You can click the Back button to change the original location of your installation.

7. **Click Next.**

   A dialog box with a progress bar appears and the bar moves all the way across a few times. Then a dialog box with a Finish button appears.

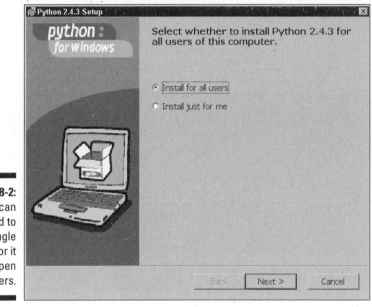

Figure 18-2:
Python can be limited to a single user or it can be open to all users.

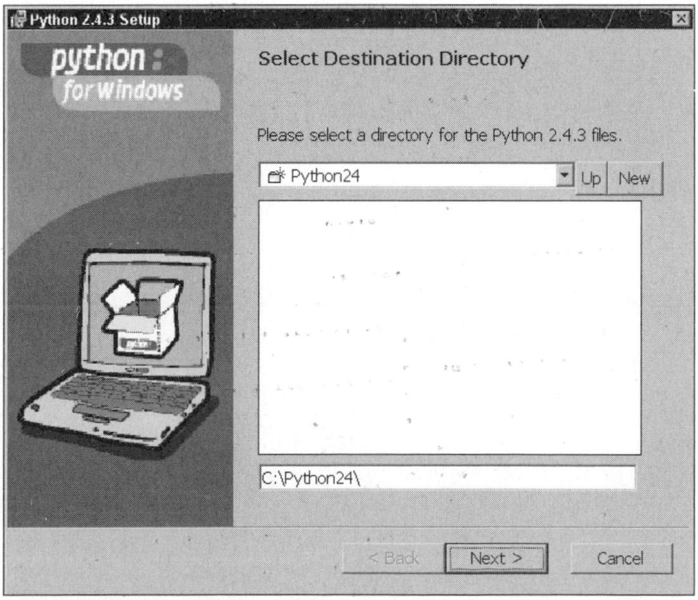

Figure 18-3:
You can install Python in any directory on your system.

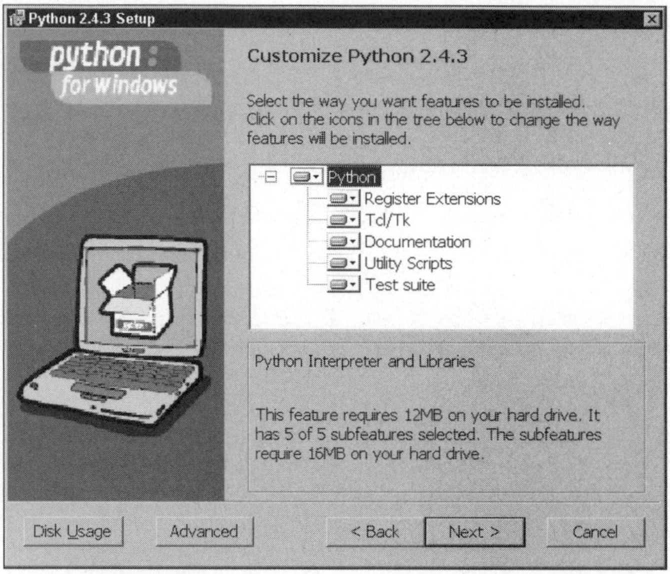

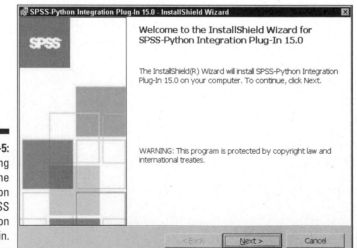

Figure 18-4:
You can exclude parts of Python from the initial installation.

8. **Click Finish.**

   The fundamentals of Python are installed. The SPSS add-ons are not. The main installation dialog box appears again.

9. **Click the Install SPSS-Python Integration Plug-in option.**

   A dialog box containing a progress bar appears and the progress bar moves across from left to right a few times, then the dialog box in Figure 18-5 appears.

Figure 18-5:
Beginning of the installation of the SPSS Python Plug-in.

10. **Click Next.**

    The freeware license agreement for the Python plug-in appears.

11. **If you agree with the terms of the license, click the I Accept the Terms option, and then click the Next button.**

    The dialog box in Figure 18-6 appears. The installation process is searching for the directory where SPSS is installed. If it doesn't find it, or if it finds the wrong one, you can correct the name of the directory.

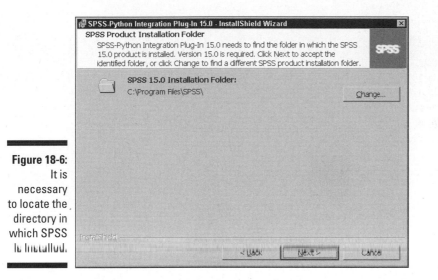

Figure 18-6:
It is necessary to locate the directory in which SPSS is installed.

12. **Click the Next button.**

    A dialog box appears showing the location of the installation of Python and SPSS.

13. **If the list of locations for the previous installations are correct, click Install.**

    If the locations are not correct, the installation is doomed to failure. Click the Cancel button so you can start over from the beginning

14. **Click Finish.**

    The process completes and the main installation window reappears.

15. **Click Exit.**

Python and the SPSS Python plug-in are now installed and ready to go to work for you. But you really need to do one more thing before you try to use Python. You must install some auxiliary modules, as described in the next section.

A Language inside a Language

Python runs from inside a Syntax program. After you have the Python plug-in installed, you can write Syntax programs and include Python programs inside them by surrounding the Python code with the correct Syntax commands.

To run a simple Python command, and check whether your plug-in is installed and working, do the following. In the main SPSS window, choose File⇨New⇨ Syntax. Then, in the Syntax Editor dialog box, enter the following three lines:

```
BEGIN PROGRAM.
print "Python speaks!"
END PROGRAM.
```

This is a Syntax program with a one-line Python program embedded inside it. Everything between the Syntax commands BEGIN PROGRAM and END PRO-GRAM is Python. In this example, the included program is one line of Python consisting of a print statement. Choosing Run⇨All on the Syntax Editor menu produces the following text as output:

```
BEGIN PROGRAM.
print "Python speaks!"
END PROGRAM.
Python speaks!
```

This output is a complete listing of the program, followed by the output from the program. I'm sure you've notice from your earlier activities that SPSS always lists the Syntax source code before running the program.

To access SPSS data and commands from inside a Python program, you must use an import statement to make SPSS available. You need only import SPSS one time in a Python program, but once you do, you have access to all data and even to Syntax language commands. For example, from inside a Python program, you can use the Syntax command LIST to output the values of all the variables of all the cases:

```
BEGIN PROGRAM.
import spss
spss.Submit("LIST.")
END PROGRAM.
```

This is a Syntax command issued from inside a Python program that is being run inside a Syntax program. I need to mention a couple of things to be care-ful about here. Notice the period at the end of the quoted Syntax command — all Syntax statements require a terminating period. Also notice that all the Python code is in lowercase, except for the capital S on Submit — Python is case-sensitive and getting the case wrong is the same as misspelling a word.

Finding out about modules

You can install Python modules that have classes and functions in them to help you with processing. In the examples in the preceding section, the spss module contains the Submit function, which is used to execute Syntax commands. The spss module also contains other useful functions. You can look at its contents with the following program:

```
BEGIN PROGRAM.
import spss
help(spss)
END PROGRAM.
```

This program uses the Python help function to output information about the module. But the information is larger than can be displayed in SPSS Viewer in a single chunk, so it's necessary to expand things to see it all. Double-click the text shown in SPSS Viewer, and the complete text (all 1769 lines of it) appears in an SPSS Text Output dialog box. You can scroll through the text to find out what's there.

You can use help to find out about almost anything having to do with Python. For example, if you want to be more specific in your search for help, you can get help on the Submit function by executing the command help (spss.Submit). You can also be more general in your search. For example, the following program gives you a complete list of available modules:

```
BEGIN PROGRAM.
help("modules")
END PROGRAM.
```

To find out about a specific module, it must be imported first. For example, the following program produces a list of functions available in the time module:

```
BEGIN PROGRAM.
import time
help(time)
END PROGRAM.
```

You can even get help on help with the following:

```
BEGIN PROGRAM.
help()
END PROGRAM.
```

Installing more modules

A large number of modules are already installed with Python, but more are available. You can find a collection of Python modules at the following Web site:

```
http://www.spss.com/devcentral/
```

You will find several Python modules at this site, but you probably don't need them all. If you find yourself with an urge to do lots of things with Python, I suggest that you take a look at the following:

✔ **spssaux:** This module contains utilities, many of which are used by other modules. Among other things, it makes it possible for you to work with SPSS definitions and produce output. It provides pathways for data coming out of SPSS to be input into Python.

✔ **spssdata:** This module provides access to the data of the current SPSS dataset. It can be instructed to fetch the data one case at a time or loop through all the cases returning data to your program.

Multiple Commands with One Submit

The Submit function can be used to execute more than one Syntax statement. You can do it using a series of Submit statements, or you can issue a series of statements with one Submit function call. The following example shows you how you can use the Submit function call with an array of command strings instead of just one:

```
BEGIN PROGRAM.
import spss
spss.Submit(["GET FILE='c:/Program Files/SPSS/Cars.sav'.",
             "PRINT / ALL.",
             "EXECUTE."])
END PROGRAM.
```

With this form, all the punctuation must be correct so that Python can figure out what you mean:

✔ The square brackets ([and]) indicate an array instead of a single quoted string.

✔ Each string inside the array has its own beginning and ending double quotes (") to delimit the beginning and ending of the string.

✔ Forward slashes (/) are used inside the path name of a file. Backward slashes have a special meaning to Python and should not be used.

✔ A string within a string is delimited by a different kind of quote. In this example, the inside string is defined by single quotes (') while the containing string is defined by double quotes (").

✔ A comma is placed between the strings. Without a comma between them, Python will combine the two strings into one.

✔ Each Syntax command is terminate with a period.

Working with SPSS Variables

You can read the values of SPSS variables and do an analysis on them inside a Python program. The spss module gives you access to them. The following example does a simple analysis using only the scale variables:

```
BEGIN PROGRAM.
import spss
spss.Submit("GET FILE='c:/Program Files/SPSS/Cars.sav'.")
varList=[]
for i in range(spss.GetVariableCount()):
    if(spss.GetVariableMeasurementLevel(i)=='scale'):
        varList.append(spss.GetVariableName(i))
if(len(varList)):
    spss.Submit("DESCRIPTIVES " + " ".join(varList) + ".")
END PROGRAM.
```

This example program performs the following actions:

✔ The spss module is imported.

✔ The spss.Submit function is called to load the information from disk.

✔ An array named varList is declared. The array is initially empty.

✔ A loop executes with the variable i ranging from 0 to the total number of variables in the loaded data set. The total number of variables is determined at the top of the loop by a call to GetVariableCount().

✔ Inside the loop, the call to GetVariableMeasurementLevel() returns a descriptor of the type of the variable. If it is a scale type, the variable name is retrieved with a call to GetVariableName() and the name is appended to the array varList[].

✔ Inside a second if statement, a call is made to len() to determine whether anything has been added to the array. If it hasn't, there are no scale variables, and no output will be produced.

✔ If at least one variable is in the array, a call is made to Submit() to execute a Syntax language DESCRIPTIVES command. The result is the output shown in Figure 18-7.

Figure 18-7:
A table
produced
from Python
executing a
Syntax
DESCRIP-
TIVES
command.

Descriptive Statistics

	N	Minimum	Maximum	Mean	Std. Deviation
Miles per Gallon	398	9	47	23.51	7.816
Engine Displacement (cu. inches)	406	4	455	194.04	105.207
Horsepower	400	46	230	104.83	38.522
Vehicle Weight (lbs.)	406	732	5140	2969.56	849.827
Time to Accelerate from 0 to 60 mph (sec)	406	8	25	15.50	2.821
Valid N (listwise)	392				

In this example, a group of variable names were entered as part of a single string. The command string looked like the following:

```
DESCRIPTIVES mpg engine horse weight accel.
```

The join() method is a Python method that accepts an array of strings and joins them as one long string with spaces inserted as separators.

Accessing SPSS from Outside

You don't have to load SPSS and use the Syntax window to run Python. Python runs on its own and you can use SPSS commands within the stand-alone Python program. The two magic words are

```
import spss
```

From the import statement in your Python programs, you will be able to call Submit() or any other function defined in the spss package. You can load other packages as you need them. You don't need to use BEGIN PROGRAM and END PROGRAM because you don't have to issue a notification of your intent to use Python.

An IDE (Integrated Development Environment) for Python provides you with almost everything you need. A Python IDE has a built-in text editor designed for the Python language, a Python runtime system, a debugger, and the ability to load modules. Several IDEs exist — use Google to search for *Python IDE* and you will find several. You will like some better than others, so don't just hang out with the first one you come across.

Chapter 19

Scripts

● ●

In This Chapter

▶ Scripting with Sax BASIC for SPSS

▶ BASIC classes and objects for SPSS

▶ Creating global and automatic scripts

● ●

*Y*ou can write BASIC language programs that run inside SPSS. Such programs are known to SPSS as scripts. SPSS has a dialog box specially designed to edit these scripts, run them, and save them to disk. In writing scripts, you have the advantage that the Sax BASIC language is common and widespread, making it easy to find documentation, both in print form and on the Internet. A good deal of documentation is also inside the SPSS help system.

Although scripts work with input data, they primarily work with output — the data displayed in SPSS Viewer. For example, you can use a script to add items to or delete items from a pivot table. You can write a script to modify a graph after it has been displayed.

Picking Up BASIC

This chapter is not a tutorial on programming using the BASIC language. You can get that information from Internet tutorials and from books on Sax BASIC and Visual BASIC. This chapter is about the particulars of using BASIC as a scripting language inside SPSS.

You should always start writing a new script by copying an old script that works. In fact, SPSS provides a number of starter scripts for you to use for this very purpose. Before you write a script of your own, you should look through the collection of scripts provided so you become familiar with what you already have. The scripts provided are complete, and one of them may perform the task you are trying to achieve.

Scripting can be used to automate some things, but it does not provide magic powers for you to do things you cannot do otherwise. All the things you can do with a script, you can also do with mouse controls. Before writing a script, step through the procedure with the mouse so you know exactly what you want the script to do.

In the SPSS documentation, only BASIC programs are referred to as scripts. Although the other languages of SPSS — Syntax and Python — fit the technical definition of scripting languages, SPSS considers only BASIC as its scripting language.

Scripting Fundamentals

Sax BASIC uses a few of the fundamental concepts of object-oriented programming. It doesn't use many, but you need to have an understanding of the little bits it does use.

Through some process that I don't quite get, object-oriented programming has the reputation of being difficult to understand. It isn't. It's easy to understand but it is clumsy to explain — sort of like describing an accordion without using your hands. But let me try.

The roads and streets are full of cars. There are many different kinds and shapes of cars, but they are all cars. That means the word *car* is a specific classification of vehicle. A car is a *class*. Fred's old, beat-up, blue 1968 Chevy is a specific car. It is an *object* of the class known as car. Every actual car is an object.

In this paragraph I made reference to Fred's car as being an object. It was only a reference; not the actual object. If you have all that, you now understand every fundamental that you need about classes, objects, and references to be able to understand object-oriented programming. If you find yourself getting confused about which is what, just remember Fred's old, beat-up, blue Chevy. That's what I do, and it works for me.

Software classes, objects, and references

You already know what a pivot table is. And you know that lots of pivot tables of different sizes and types exist, but they are all pivot tables. That makes a pivot table a classification — or, in programming terms, a *class*. A specific pivot table is an *object*.

In SPSS scripts, a pivot table is an object of the class named `PivotTable`. You can't copy an entire pivot table into your program, but you can get a *reference* to it. You can think of the reference as a kind of address that can be used to access the pivot table when you want to refer to it. In your script, you can create a reference to a pivot table with a statement like the following:

```
Dim pt as PivotTable
```

In this statement, the `pt` variable is created as a reference to an object of the class named `PivotTable`. The class name, `PivotTable`, is already defined for you by SPSS. Class names are already defined for charts, documents, data cells, and several other things. (I have included a complete list in the next section as Table 19-1.) The reason the silly word `Dim` is used to declare a variable has to do with the history of the BASIC language. I chose `pt` to be the name of the reference for no particular reason. You can choose any name you like. The names used for references in the example programs supplied by SPSS are made by sticking `obj` in front of the class name, like the following:

```
Dim objPivotTable as PivotTable
```

A reference declared this way does *not* refer to an actual pivot table. Yet. You have to select a pivot table and initialize your new variable with its address.

Some classes are built into Sax BASIC, and you will encounter them in the sample scripts. For example, a class named `String` is used to declare string variables like the following:

```
Dim mystring as String
```

Or you can define the reference to an `Integer` like the following:

```
Dim myinteger as Integer
```

The classes of SPSS

A number of classes are defined and ready for you to use in your program, as listed in Table 19-1. All names of all classes, with the exception of `PivotTable`, begin with an uppercase *I*. All the references in the example programs begin with lowercase letters.

One member of the list is special. The reference name `objSpssApp`, which is of the class `ISpssApp`, has already been declared and initialized. It is ready to go in every program and acts as your access point to objects in all the other classes. By using the properties and methods of `objSpssApp`, you can acquire objects in all the other classes.

Table 19-1	SPSS Classes You Can Use in Your Program	
Class Name	*What an Object of This Class Refers To*	*Name Used in the Example Programs for References to This Class*
PivotTable	Pivot table	objPivotTable
ISpssApp	Entire SPSS application	objSpssApp
ISpssChart	Chart or graph	objSPSSChart
ISpssDataCells	Data cells	objDataCells
ISpssDataDoc	Data document	objDataDoc
ISpssDimension	Dimension	objDimension
ISpssDocuments	Documents	objDocuments
ISpssFootnotes	Footnotes	objFootnotes
ISpssInfo	SPSS file information	objSpssInfo
ISpssItem	Output item	objOutputItem
ISpssItems	Collection of output items	objOutputItems
ISpssLabels	Row or column labels	objColumnLabels and objRowLabels
ISpssLayerLabels	Layer labels	objLayerLabels
ISpssOptions	SPSS options	objSpssOptions
ISpssOutputDoc	Viewer document	objOutputDoc
ISpssPrintOptions	Printer options	objPrintOptions
ISpssPivotMgr	Pivot manager	objPivotMgr
ISpssRtf	Text	objSPSSText
ISpssSyntaxDoc	Syntax document	objSyntaxDoc

Properties and methods

Each class has a unique set of properties and methods by which you can access internal information. A *property* is a variable that is part of the class definition. Each object of a class has its own set of values for its properties.

Each property has read and write permissions. Your program can use some properties only to read values from the object, other properties to write values into the object, and still other properties for both. *Methods* are procedures associated with the object, making it possible for you to execute a set of instructions associated with the object.

To be able to do anything with an object, you need to know which properties and methods are available. You can find out about any particular class definition by using the following steps:

1. **Choose File⇨Open⇨Scripts**

   This opens the dialog box used to edit scripts. Another dialog box, named Use Starter Script, also appears but you can close it because you won't be using it here.

2. **Choose Help⇨Objects.**

   The dialog box shown in Figure 19-1 appears, showing you the relationship among classes as clickable buttons.

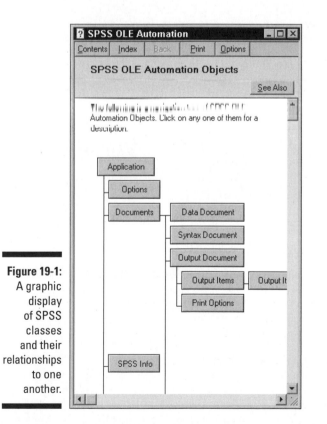

Figure 19-1: A graphic display of SPSS classes and their relationships to one another.

3. **Click the button representing the class you want to know about.**

   A window appears with a brief description of the class and example code showing how to declare a reference and how to initialize the reference with a specific object.

4. **Click Properties or Methods to get more information.**

   You are presented with a list of either property or method names.

5. **Select the name from the list and then click Display.**

   A full description of the property or method appears, along with the syntax of the code you can use to access it.

Creating a New Script

The first step in creating a script is to choose File⊅New⊅Script. A dialog box appears with the seed of a script. The seed consists only of the opening line `Sub Main` and the closing line `End Sub`. But that's not all the help you get. You also get a User Starter Script dialog box like the one shown in Figure 19-2 containing a list of scripts you can use for starters.

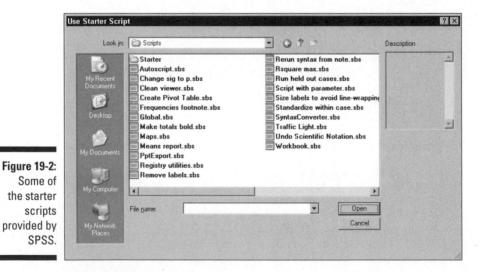

Figure 19-2: Some of the starter scripts provided by SPSS.

You may have to browse around a bit to find the script that is most like the one you want to finally produce. These are not tiny scripts. They all have several lines of code and are filled with comments explaining how they work and how you might want to change them to make them do what you'd like.

After you have found a script you want to use for your starter, save it immediately under a new name. You don't want to save it under the same name because your changed script will overwrite the starter script and you won't be able to get it back if you need it.

While you are editing a script, you need to save it to disk from time to time for safety, and then save it again when you're finished. Your script file can be stored anywhere, but it should have the suffix `.sbs` (or `.SBS`, case doesn't matter) so you will be able to load it into SPSS and use it again.

Global Procedures

You can write a procedure or a function and store it in such a way that you can run it from any other script you write. To do this, edit the global script file `global.sbs` and include your procedure in it. You can bring up the global script file for editing by clicking the number 2, which appears as a tab on the left side of the dialog box. (The tab numbered 1 takes you back to the editing window containing the script you're working on.) If you click the number 2, you see the text of some global procedures that have already been defined. All you have to do is add yours to the list.

Only one global procedures file exists, so it must contain all global procedures. You can change the file used by choosing Edit⇨Options⇨Scripts from the main SPSS window. The result may not be what you want because only one file at a time is used to contain global procedures, and changing to another file disables all the procedures now defined.

Automatic Scripts

You can set scripts to execute automatically on data triggers. To do this, the script must be included in the `autoscripts` file and enabled. The default name of the file is `autoscripts.sbs`, but you can change it to any name you like.

To control `autoscripts`, on the main window of SPSS choose Edit⇨Options and then click the Scripts tab. The dialog box shown in Figure 19-3 appears.

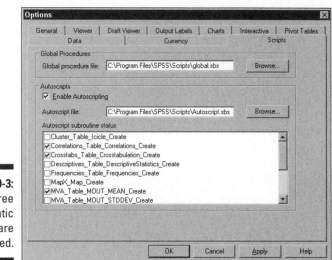

Figure 19-3:
Three
automatic
scripts are
enabled.

The name of the script determines the action that will trigger an autoscript to execute. If you plan on writing a script to execute automatically, you should browse through the scripts in the `autoscripts.sbs` file to see how they are named. The following events can be made to activate an automatic script:

✓ **Creation of a pivot table:** The name of the procedure identifies the type of table and the procedure used to create it, followed by the word `Create`, such as `Correlations_Table_Correlations_Create`.

✓ **Creation of a title:** The name of the procedure identifies the procedure used to create the title, such as `Correlations_Title_Create`.

✓ **Creation of notes:** The name of the procedure identifies the procedure use to create the notes, such as `Correlations_Notes_Create`.

✓ **Creation of warnings:** You can begin the name with the name of the procedure producing the warning, or just have the script triggered on all warnings, such as `Warning_Create`.

Part VII
The Part of Tens

The 5th Wave   By Rich Tennant

HELP DESK

TECHNICAL   ETHICAL

In this part . . .

This part describes ten modules that can be added on to SPSS, and ten places you can go on the Internet to find useful information.

Chapter 20

Ten Modules You Can Add to SPSS

SPSS comes in the form of a base system and several modules you can acquire to add on to it. If you have installed the full system, you already have a lot of these add-ons. Most are integrated and look like they are part of the base system. Some will be of no interest to you, but others could become indispensable. This chapter introduces you to them and describes what they do, but you need to refer to the documentation that comes with them for a full tutorial.

All but one of the add-ons listed here come directly from SPSS, and you can find out more about them at the SPSS Web site (`http://www.spss.com`). All of the add-ons are available in English. Some are also available in Japanese, French, German, Italian, Spanish, Chinese, Polish, Korean, or Russian.

Geoset Manager

SPSS can load map files and display them with your data in place. A number of maps are supplied as part of the SPSS base system and more are available for purchase at `http://www.spss.com`, but if you really want to get into map making, you need the Geoset Manager. It installs into the same installation directory as SPSS; its executable file is named `geosetmanager40.exe`.

Using the Geoset Manager, you can design your own maps and overlay geographical features onto an existing map. Maps can be made up of layers, with each layer displaying certain features. If you want to customize your maps before you add your data to them, you need to use Geoset Manager.

The Geoset Manager comes on the CD along with the base SPSS, so you will only need to have an authorization code that unlocks it so you can use it.

Amos

Amos is an interactive interface that you can use to build structural equation models. Using the path diagrams you create with Amos, you can discover unexpected relationships and gain more insight into the meaning of your data.

Amos provides a more intuitive interface than plain SPSS for a certain family of problems. Amos contains structural modeling software that you control with a drag-and-drop interface. Because the interface is intuitive, you can create models that come closer to the real world than the multivariate statistical methods of SPSS. You set up your variables and then you can perform analyses using hypothetical relationships.

Amos enables you to build models that more realistically reflect complex relationships with the ability to use observed variables, such as survey data or latent variables like "satisfaction" to predict any other numeric variable. Structural equation modeling, sometimes called path analysis, helps you gain additional insight into causal models and the strength of variable relationships.

Missing Value Analysis

The missing values in your data (whether or not they are excluded from your calculations) will have an effect on the outcome. The SPSS Missing Value Analysis add-on can let you know in what way your pattern is affecting your results.

With this add-on, you can detect patterns of missing data. Armed with this information, you can determine the cause of the missing information or you can use regression or expectation algorithms to generate values. By properly managing missing data, you can use all your data instead of limiting analysis to complete cases. Handling missing data wisely can remove hidden bias.

SPSS Missing Value Analysis can tell you whether you have a serious missing data problem. You can find this out through the data patterns report, which is a case-by-case overview displaying the extent and overview of the missing data.

The missing data information can be used to improve survey questions that you identify as possibly troublesome or confusing. You can detect whether there is a relationship between missing values — values missing for one variable could be related to missing variables of another.

Regression Models

The Regression Models add-on plugs directly into the SPSS base software and provides a larger selection of statistical analysis methods. It includes some additional diagnostic capabilities. You can predict group membership in key groups. You could, for example, build a model that will predict which product a customer is most likely to order.

With this add-on, you can regress a categorical variable with multiple categories based on a set of independent variables. You would be able to analyze using forward entry and backward elimination, and move in steps forward or backward. This process exposes the most accurate predictors. If you wind up with a large number of predictor variables, you can use them together to come up with a more accurate result. Using this add-on, you can group people according to their predicted actions.

Advanced Multivariate Models

The Advanced Multivariate Models add-on specializes in complex relationships among multiple variables. The procedures are more sophisticated than the multivariate algorithms found in the base SPSS system and can be used to produce more dependable conclusions. This is a set of univariate and multivariate analysis techniques that you can apply to real-world problems.

In addition to the general linear models and mixed models, SPSS Advanced Models now includes procedures for Generalized Linear Models (GLMs) and Generalized Estimating Equations (GEEs).

The GLMs include linear regression for normally distributed responses, logistic models for binary data, and log linear models for count data. The GEEs extend generalized linear models to accommodate correlated longitudinal data and clustered data. The SPSS Advanced Model also includes General Linear Models (GLM) and Hierarchical Linear Models (HLM).

Exact Tests

The SPSS Exact Tests add-on makes it possible to be more accurate in your analysis of small data sets and data sets that contain rare occurrences. It gives you the tools you need to analyze such data conditions with more accuracy than would otherwise be possible.

When only a small sample size is available, this add-on enables you to use the smaller sample and have more confidence in the results. The purpose is to make it possible for you to perform more analyses in a shorter period of time. This add-on allows you to conduct different surveys rather than spend time gathering samples to enlarge the base of the surveys you have.

The processes you use, and the forms of the results, are the same as those in the base SPSS system, but the internal algorithms are tuned to work with smaller data sets. The Exact Tests add-on provides more than 30 tests covering all the nonparametric and categorical tests you normally use for larger data sets. Included are one-sample, two-sample, and K-Sample tests with independent or related samples, goodness-of-fit tests, tests of independence, and measures of association.

SPSS Categories

The SPSS Categories add-on is designed for you to reveal relationships among your categorical data. To help you understand your data, SPSS Categories uses perceptual mapping, optimal scaling, preference scaling, and dimension reduction. These techniques make it possible to visually interpret the ways in which your rows and columns relate to one another.

SPSS Categories performs its analysis and displays results so you can understand ordinal and nominal data. It uses procedures similar to conventional regression, principal components, and canonical correlation. It performs regression using nominal or ordinal categorical predictor or outcome variables.

The procedures of SPSS Categories make it possible to perform statistical operations on categorical data. The scaling procedures can be used to assign units of measurement and zero-points to your categorical data, which makes it possible to access new groups of statistical functions by allowing you to perform analyses on variables based on mixed measurement levels. You can use correspondence analysis to help you numerically evaluate similarities between two or more nominal variables and to summarize your data according to components you select. You can collect variables of different measurement levels into sets of their own, and then analyze the sets by using nonlinear canonical correlation analysis.

This add-on can be used to produce perceptual maps and biplots. Perceptual maps are high-resolution summary charts graphically displaying similar variables or categories that are close to one another. Perceptual maps give you insights into relationships between more than two categorical variables. Biplots make it possible to look at the relationships among cases, variables, and categories so you can see how they relate.

SPSS Trends

SPSS Trends can be used to quickly construct expert time-series forecasts. It includes statistical algorithms you can use to analyze historical data and predict trends. You can set it up to analyze hundreds of different time series at once instead of running a separate procedure for each one.

The software is designed to handle the special situations that arise in trend analysis. It automatically determines the best-fitting ARIMA (Autoregressive Integrated Moving Average) or smoothing model. It automatically tests data for seasonality, intermittency, and missing values. The software detects outliers and prevents them from unduly influencing the results. The graphs generated include confidence intervals and indicate the model's goodness of fit.

As you gain experience at forecasting, SPSS Trends will allow you control over every parameter when building your data model. You can use the Expert Modeler in SPSS Trends to recommend a starting point or to check calculations you've done by hand.

You can design models and save them in such a way that your forecasts can be updated on the arrival of changed data, or new data, without the necessity of re-estimating the model. Also, you can write scripts to update the models as situations change.

SPSS Map

The SPSS Map add-on consists of map data files that you can use to produce graphic geographic output of the distribution of your data. Using the sample maps supplied with the base SPSS system, the techniques for using SPSS Maps are described in Chapter 12.

This add-on includes detailed maps from all over the world along with numerous sample data sets that can be used to display map data.

SPSS Complex Samples

The SPSS Complex Samples module is for working with complex sampling methods such as stratified, clustered, or multistage sampling.

Stratified sampling is choosing to sample within subgroups of the survey population. For example, subgroups might be a specific number of males or females or contain people in certain job categories or people of a certain age group.

Clustered sampling is sampling from groups of sampling units. Clusters can include schools, hospitals, or geographic areas, with sampling units that might be students, patients, or citizens. Clustering often helps makes surveys more cost effective.

Multistage sampling is selecting a first-stage sample based on groups of elements in the population, and then creating a second-stage sample by drawing a subsample from each selected unit in the first-stage sample. This process can be repeated to select higher-stage samples. For example, in a face-to-face survey, you might sample individuals within households and city blocks.

The software allows you to incorporate the sampling design into your survey analysis. It can work more accurately with numerical and categorical outcomes within these complex sample designs by using its unique algorithms for analysis and prediction.

From the start, the Sampling Wizard can be used to describe the data gathering scheme. You can create plans or describe existing plans, and then analyze the data to produce results.

As your output, you can include public-use data sets that include your analysis plans. This output makes it possible to later plug data into the plans to extend the analysis with new information.

Chapter 21

Ten Useful Things You Can Find on the Internet

*T*he SPSS system is used in enough places and by enough people that it appears for various reasons on the Internet. Some of the Web pages are produced by the same company that manufactures the software, but many pages are produced by others who are interested in using SPSS. This chapter gives you a general idea of the purpose of some of the most useful sites.

You may not want to type the URLs in this chapter, so I created a Web page to make it possible for you to simply click the links. Go to the following address:

```
www.dummies.com/go/spss
```

I also maintain a Web page at the following address:

```
www.belugalake.com/spss
```

SPSS Humor

You will find an amazing variety of SPSS stuff on the Internet, from specific programming to general commentary. Even humor. The following two Web sites are dedicated to SPSS and statistics jokes:

```
http://www.ilstu.edu/%7egcramsey/Gallery.html
http://www.kingdouglas.com/SPSS/DiverseCultures/Humor.htm
```

The SPSS Home Page

The Web site of the SPSS company, and the Web site from which you can locate articles, programs, add-ons, and general news about SPSS, can be located by pointing your Web browser to the following address:

```
http://www.spss.com
```

Another way to get to the same Web site is to use the menus of SPSS and choose Help⇨SPSS Home Page.

From this base Web site, you can locate the SPSS home page for 34 countries other than the United States. This page allows you to specify a search string so you can locate the article, training service, or detailed description of whatever you want. The Web site is quite large and will probably contain some information about whatever it is you are trying to research, whether it's statistics in general or SPSS in particular.

SPSS Developer Center

Whether you want to write SPSS programs or become otherwise knowledgeable about the workings of SPSS, you will need to check out the developer center. It has information on all sorts of SPSS programming. You can find the center here:

```
http://www.spss.com/devcentral/
```

You can use your Web browser to go directly to this site, or you can use the menu on the main window of SPSS and choose Help⇨SPSS Developer Central. Plenty of information is on that Web site, so you will need to browse around to find what you're looking for.

You can download utility programs already written and ready to go. You can also download graphics examples and new statistical modules, and you will find a large number of articles on the inner workings of SPSS technology.

SPSS has forums where you can interact with people inside the SPSS company and with other SPSS users. If you have a question or a problem, this Web site is a good place to bring it.

User Groups

SPSS has experts and experienced users, and a lot of them are ready to answer questions. If you have a question, instead of sitting there with a giant question mark floating over your head, check out these sites:

```
http://www.spssusers.co.uk/
http://www.spsslog.com/
```

Mailing Lists and News Groups

A surprisingly large number of mailing lists are based on statistics. If you want, you can join a mailing list and receive copies of the ongoing discussions. You need not make your presence known until you have a question or have something to contribute. You can choose from among the mailing lists at the following sites:

```
http://listserv.uark.edu/archives/ua-spss-user-group.html
http://list.haifa.ac.il/mailman/listinfo/spss-users
http://www.stattransfer.com/lists.html
```

The following is a newsgroup frequented by SPSS users:

```
comp.soft-sys.stat.spss
```

To take a look at examples of newsgroup postings, you can read the archived articles at the following location:

```
http://groups.google.com/group/comp.soft-sys.stat.spss
```

For statistics in general, three newsgroups exist. Following are the name and URL for the archived Web site of each one. You can look at the archives and get an idea of the type and frequency of posts:

```
sci.stat.consult
http://groups.google.com/group/sci.stat.consult

sci.stat.edu
http://groups.google.com/group/sci.stat.edu

sci.stat.math
http://groups.google.com/group/sci.stat.math
```

Python Programming

This book gives you a small peek at the things you can do with Python. It's true that it's a language built into SPSS, but it's much more than that. It's much more than you will ever need. Python is a general-purpose programming language like C or Java — that is, you can use it to do anything you might ever want to do.

And it runs almost anywhere. You will find versions of Python for Linux, Windows, Apple, and even cell phones. That's right. You can use it to program your cell phone.

If you want to go further into Python, there is no better place to start than the Python Language Web site. Lots of stuff is there, but two things are of prime importance: complete documentation (tutorials, examples, and more) and a free copy of Python that you can download and install on your machine:

```
http://www.python.org
```

Quite often, newcomers to programming find themselves put off by the terms used to describe the programming language. Don't be. It's a lot easier to understand the fundamentals of programming than it is to understand statistics; it's just that nerds like to show off by talking that way. I know, I've done it myself.

The Python Web site is friendlier than most of its kind, and it could be used as an excellent place to start learning how to program. Programming is not a bad hobby, but it can be habit forming. Be careful — you can find that happening to you, and before you know it, you're on the road to becoming a nerd.

The following Web sites are helpful when programming Python within SPSS:

```
http://www.nettakeaway.com/tp/
http://www.american.edu/econ/pytrix/pytrix.htm
http://www.spss.com/devcentral
```

Script and Syntax Programming

You can find programs and programming tutorials for the various SPSS languages. All the Web sites listed concern themselves with programming SPSS. Most have commentary and suggestions along with programs, some are tutorials on programming, and some have programs that you can download and use.

Syntax language:

```
http://flash.lakeheadu.ca/~boconno2/boconnor.html
http://www.ats.ucla.edu/stat/spss/seminars/spss_syntax/
http://bama.ua.edu/~jhartman/689/syntax.ppt
http://www.hmdc.harvard.edu/pub_files/SPSS_Syntax.pdf
http://www.sharewareconnection.com/easysyntax.htm
http://oit.utk.edu/scc/HowToUseSPSSSyntaxFilesOnUNIX.pdf
http://www.longitudinal.stir.ac.uk/SPSS_support.html
http://www.unt.edu/rss/class/spssclass1/Examples.htm
http://www.healthinformation.on.ca/spss.html
http://www.socio.com/helpdata1.htm
```

Scripts (Sax BASIC):

```
http://pages.infinit.net/rlevesqu/SampleScripts.htm
http://www.xs4all.nl/~jhckx/spss/scripts/
http://ttsoftware.com/manuals/basic32.pdf
http://www.ocair.org/files/VBAwksp/spss.htm
http://www.freedomscientific.com/fs_support/
         BulletinView.cfm?QC=426
http://www.spssusers.co.uk/Tips/saxbasic_doc.html
```

General SPSS programming:

```
http://www.spsstools.net/
http://www.spss.com/downloads/Papers.cfm?List=all&Name=all
http://scripts.filehungry.com/product/java/javabeans/
         development_tools/java_spss_writer
```

Tutorials for SPSS and Statistics

One of the things the Web does very well is present tutorials. In fact, that is what it was originally designed to do — instead of the advertising and marketing arena that it has become. This section contains a short list of tutorial Web sites, but there are certainly more. Some of the sites are for statistics, some are for SPSS, and some are for both.

If you are looking for a tutorial, you will probably need to search through several of these sites to find the one you want to start with. Some are better than others. They all emphasize certain characteristics and capabilities of the software. Some specialize in statistics for a particular subject, which may or may not be to your advantage. Some were designed using older versions of SPSS, but the capabilities of SPSS have expanded, not contracted, so those lessons should still be valid.

This list is only a small percentage of the total. These are for general-purpose studies, but some sites become specific in the types of statistics they present. If you wanted to narrow your search to say, medical statistics, you could enter the search string *SPSS tutorial medical* to turn up a number of specialized sites.

SPSS tutorials:

```
http://www.hmdc.harvard.edu/projects/SPSS_Tutorial/
          spsstut.shtml
http://calcnet.mth.cmich.edu/org/spss/toc.htm
http://www.ats.ucla.edu/STAT/spss/
http://cs.furman.edu/rushing/mellonj/spss1.htm
http://www.utexas.edu/its/rc/tutorials/stat/spss/spss1/
http://www.shef.ac.uk/scharr/spss/
http://www.students.stir.ac.uk/docs/spss/spss.html
http://www.datastep.com/SPSSTraining.html
http://www.stat.tamu.edu/spss.php
http://www.uri.edu/ois/iits/research/spss/spss75.htm
http://academic.uofs.edu/department/psych/methods/
          cannon99/spssmain.html
http://its.unm.edu/introductions/Spss_tutorial/
```

General statistics tutorials:

```
http://www2.chass.ncsu.edu/garson/pa765/statnote.htm
http://www.meandeviation.com/tutorials/stats/
http://davidmlane.com/hyperstat/
http://www.psych.utoronto.ca/courses/c1/statstoc.htm
http://www.statsoft.com/textbook/stathome.html
http://mail.pittstate.edu/~winters/tutorial/
http://math.about.com/od/statistics/Statistics_
          Tutorials_and_Resources.htm
```

Sites for both SPSS and general statistics:

```
http://www.utexas.edu/its/rc/tutorials/
http://www.psych.utoronto.ca/courses/c1/Welcome.htm
http://www.uni.edu/its/us/document/stats/spss2.html
http://www.tulane.edu/~panda2/Analysis2/ahome.html
http://web.uccs.edu/lbecker/SPSS/content.htm
http://www.cas.lancs.ac.uk/short_courses/intro_spss.html
```

SPSS Wiki

A *Wiki* is a Web site with documents that are constantly updated. You can join as a reader and as a contributor. The SPSS Wiki acts both as a reference source and as a workbook for SPSS statistical procedures. It can be used equally well by both novices and experts.

Instructions on the Web page tell you how to use the Wiki to find what you're looking for and how to contribute to the constantly growing body of information. You will find the SPSS Wiki at the following location:

```
http://spss.wikia.com
```

PSPP, a Free SPSS

You have probably heard of the Free Software Foundation and GNU. The members are involved in developing open source software (that means free, to me and you). The PSPP project is developing an SPSS workalike. It's not possible for me to say how much has been finished and tested, because that changes almost daily, but claims are being made that it supports a large subset of SPSS. Its statistical procedure support is limited but growing.

I'm not recommending it, but I'm not poo-pooing it either. If you're interested, you can download a copy and try it for yourself. It can be downloaded in different forms and in different ways. You can find out all about how to do that at this Web site:

```
http://www.gnu.org/software/pspp/
```

You can get the latest stable version or you can get a copy of the current version while it's under development. I recommend that you get the latest stable version, at least to begin with, unless you are either a programmer or love surprises.

Besides the normal descriptive text found on the Web site, you will find e-mail addresses and IRC channels for discussions and support. You can register to be notified of future releases.

Glossary

Analysis of covariance: See *ANCOVA*.

Analysis of variance: See *ANOVA*.

ANCOVA: Analysis of covariance. ANOVA with the addition of a second or third covariate.

ANOVA: Analysis of variance. Using an F-ratio to test the fit of a linear model.

ascending: A sorting order. The values range in order from small to large. See also *descending*.

autoscript: A script that executes automatically in response to the output of certain data. See also *script*.

average: The result of adding several values and then dividing by the number of values. See also *mean* and *mode*.

base: The main system of SPSS. Modules can be added to expand SPSS, but the base system is always present.

BASIC: See *script*.

binning: The process of dividing the values of a variable into groups. Each group is a range of values and can be thought of as being sorted into its own bin. This is also called clustering.

bivariate: Two variables.

break variable: When organizing data into tabular form, the break variable is used to group the information. At the point in the report where the break variable changes value, a subtotal line is generated, or a new page is started, or some other break appears in the report.

case: Any single group (or row) of constant values. All the values in a single row. It is also called a single record.

case summary: A simple table that directly summarizes values from the cases.

categorical variable: A type of variable that can take on only one of a specific set of values, such as year of birth, make of car, or favorite color. See also *scale, ordinal, nominal, dichotomy,* and *binning.*

category: A possible value of a categorical variable.

chart: See *graph.*

clustering: See *binning.*

coefficient of determination: A statistic used to determine the correctness of the fit of regression coefficients.

command language: See *Syntax.*

confidence interval: A range around an average into which a specified percentage of the values appears. For example, if gravel trucks for a company deliver an average of 190 loads per month, but 95% of the trucks deliver between 183 and 194 loads, the 95% confidence interval ranges from a low of 7 below to a high of 4 above.

constant: A number. See also *variable.*

correlation: The degree of similarity or difference between two variables.

covariance: A comparison of the variance of one set of values with that of another.

covariate: A variable that takes part in the prediction of an outcome. An independent variable in regression. It is secondary to the relationship of the main independent variable.

cutpoint: A number used as a divider to split values into groups, as in binning.

data set: The data displayed in the Data Editor window, whether you loaded it from a file, entered it from the keyboard, or both. Multiple data sets can be loaded and will appear in separate windows. They will be labeled DataSet1, DataSet2, and so on.

degrees of freedom: The minimum number of values that must be specified to determine all the data points. This number is usually one less than the number of values used in the calculation.

delimiter: A character used to indicate the beginning of, ending of, or separation between individual values in a series of strings of characters. For example, the string of characters 59,21,34 is a series of comma-delimited numbers.

dependent variable: A variable that is compared against one or more other variables. Also called a predicted variable. See also *independent variable*.

descending: A sorting order. The values range in order from large to small. See also *ascending*.

deviation: The amount by which a measurement differs from some fixed value.

dichotomy: A variable with only two possible values, such as yes/no, true/false, or like/dislike. It is a specific type of categorical variable. See also *categorical variable*.

dodging: Plotting points on a graph so they appear next to one another instead of one of top of the other.

error: Two kinds of errors exist in the world of statistics. The conventional kind comes about when you do something wrong and get a bogus result. The other kind is calculated — that is, you figure the amount of error present in the results you get from the data you have. With modern survey techniques, you will often hear the term "margin of error" for this second type.

faceting: See *paneling*.

F-ratio: A comparison of the variance of unexpected values with the variance of expected values.

frequency distribution: The collection of values that a variable takes in a sample.

geoset: A file containing map information in a format that can be used for display and annotation by SPSS.

GLM: General Linear Model. A general procedure for analyzing variance, covariance, and regression.

graph: A non-numeric display of values. The terms graph and chart are used in SPSS internal documentation almost interchangeably.

GUI: Graphical user interface. Control of an application with windows and a mouse.

histogram: A graphical display of a distribution in which the extent of each rectangle represents the magnitude (as in a bar chart) and the width of each rectangle represents the magnitude of the bin. The area of each rectangle thus represents the frequency.

hoc: See *post hoc.*

independent variable: A variable whose values are used as the basis of a comparison. See also *dependent variable.*

kurtosis: A measure of the peakedness of the bell curve. A positive number indicates more of a peak than standard; a negative number indicates flatness of the line.

Levene test: A test to determine whether the variance of two groups is significantly different or significantly the same.

linear: A straight line. No curves.

mean: 1. Another word for average. 2. A calculated value equally distant from the two extreme values. 3. The temperament of the person making you learn this stuff. See also *average* and *mode.*

missing data: If you declare a value for a variable as representing the fact that no value is present, the missing value will not be included in calculations.

mode: The value that occurs most frequently in a given set of data. See also *average* and *mean.*

module: A utility that can be added to SPSS.

multiple response set: A special variable that has its content generated from the content of two or more other variables. In SPSS, it doesn't appear in the Data View (in the Data Editor window), but does appears when you select variable names for other activities.

multivariate: Multiple variables.

nominal: Numbers that specify categories are nominal. For example, yes, no, and undecided could be represented by 2, 1, and 0. See also *scale, ordinal,* and *categorical.*

nonlinear: Not in a straight line. Curved.

OLAP cubes: Online Analytical Processing cubes. A multilevel table containing totals, means, or some other statistic in which each level of the table contains the values relating to one value of a categorical variable.

Online analytical processing: See *OLAP.*

ordinal: Types of numbers that specify the order of occurrences. In English, the ordinal forms of 1, 2, and 3 are first, second, and third. See also *scale, nominal,* and *categorical.*

outliers: The extreme values of a variable. Generally, they are the five largest and five smallest values.

paneling: Adding another dimension of data to a graphic display causing the layout to be replicated a number of times to accommodate the values of the data along the new dimension. This process is also known as faceting.

Pearson's Product Moment Correlation: Commonly call Pearson's correlation. It represents the degree of linear relationship between two variables.

periodicity: The interval of repetition at which data recordings are made.

pivot table: A table with names identifying the rows and columns. Swapping the rows and columns to make the table appear in a different form, but containing the same data, is known as pivoting the table. The tables in SPSS Viewer are pivot tables.

post hoc: Cause and effect — some condition arises as the result of a previous condition.

p-p plot: A proportion-proportion plot. The observed cumulative proportion is plotted against the expected cumulative proportion.

predicted variable: See *dependent variable.*

probit: A nonlinear function of probability.

pyramid: A special form of a histogram where the bars representing the value extend to the sides from a center line. It often assumes the shape of a pyramid.

Python: A general-purpose programming language that can also be used to program SPSS internal operations.

q-q plot: A quantile-quantile plot. The quantiles of the observed values are plotted against the quantiles of a specified distribution.

quantiles: A set of values chosen to divide a sampling of data into groups, each containing (as far as possible) an equal number of values.

quartile: Specific values that divide all the values into four groups, with an equal number of values in each group. The groups are generally called the first, second, third, and fourth quartiles.

R: See *coefficient of determination.*

recoding: The conversion of a set of values to a new set of values. For example, if you have yes/no coded as 0/1, you can recode the values to 1/2.

record: Any single collection of values for the variables defined in SPSS. A record is all the values of a single row. It is a single case or row.

regression: Determining the "best fit" equation for the relationship between two variables. See also *dependent variable* and *independent variable.*

row: Any single collection of values for the variables defined in SPSS. It appears as a single row in the Data View window. It is a single case.

scale: A type of number that uses a standard by which something is measured, such as inches, pounds, dollars, or hours. See also *ordinal, nominal,* and *categorical.*

script: A program written in the BASIC language. It is different than Syntax and Python.

skewness: A measure of the unevenness of the distribution of data. Positive skewness indicates more high values, while negative skewness indicates more low values.

SPSS: Statistical Package for the Social Sciences.

standard deviation: A calculated indicator of the extent of deviation for a specific collection of data. The value is derived from the variations when the points are compared to a standard bell-shaped curve. It is the square root of the variance.

standard error: A measurement of the magnitude of the change from one sample to the next.

statistic: A single number calculated in a specific way. Some examples of types of a statistics are sum, mean, deviation, and average.

statistics: A collection of statistical values.

string: A series of characters making up a name or even a complete sentence. Quite often the beginning and ending of a string is delimited by quotes.

Syntax: The name of the programming language fundamental to SPSS. All actions performed by SPSS are in response to the internal interpretation of Syntax commands. In the SPSS documentation, Syntax is sometimes referred to as the command language.

t: The number of degrees of freedom. A continuous distribution with density symmetrical around the null value and a bell-shaped curve.

thematic map: A geographical map as displayed by SPSS listing statistical data for each named area.

univariate: A statistic derived from the values of one variable. Examples are mean, standard deviation, and sum.

variable: A place to store constants. A variable can store a number of constants (one for each case). Each case (or row) in SPSS consists of a collection of constant values assigned to variables.

variance: The average of the differences between a set of measured values and a set of expected values on a standard bell-shaped curve. It is the square of the *standard deviation*.

Index

BUSINESS, CAREERS & PERSONAL FINANCE

0-7645-9847-3

0-7645-2431-3

Also available:
- Business Plans Kit For Dummies
  0-7645-9794-9
- Economics For Dummies
  0-7645-5726-2
- Grant Writing For Dummies
  0-7645-8416-2
- Home Buying For Dummies
  0-7645-5331-3
- Managing For Dummies
  0-7645-1771-6
- Marketing For Dummies
  0-7645-5600-2

- Personal Finance For Dummies
  0-7645-2590-5*
- Resumes For Dummies
  0-7645-5471-9
- Selling For Dummies
  0-7645-5363-1
- Six Sigma For Dummies
  0-7645-6798-5
- Small Business Kit For Dummies
  0-7645-5984-2
- Starting an eBay Business For Dummies
  0-7645-6924-4
- Your Dream Career For Dummies
  0-7645-9795-7

HOME & BUSINESS COMPUTER BASICS

0-470-05432-8

0-471-75421-8

Also available:
- Cleaning Windows Vista For Dummies
  0-471-78293-9
- Excel 2007 For Dummies
  0-470-03737-7
- Mac OS X Tiger For Dummies
  0-7645-7675-5
- MacBook For Dummies
  0-470-04859-X
- Macs For Dummies
  0-470-04849-2
- Office 2007 For Dummies
  0-470-00923-3

- Outlook 2007 For Dummies
  0-470-03830-6
- PCs For Dummies
  0-7645-8958-X
- Salesforce.com For Dummies
  0-470-04893-X
- Upgrading & Fixing Laptops For Dummies
  0-7645-8959-8
- Word 2007 For Dummies
  0-470-03658-3
- Quicken 2007 For Dummies
  0-470-04600-7

FOOD, HOME, GARDEN, HOBBIES, MUSIC & PETS

0-7645-8404-9

0-7645-9904-6

Also available:
- Candy Making For Dummies
  0-7645-9734-5
- Card Games For Dummies
  0-7645-9910-0
- Crocheting For Dummies
  0-7645-4151-X
- Dog Training For Dummies
  0-7645-8418-9
- Healthy Carb Cookbook For Dummies
  0-7645-8476-6
- Home Maintenance For Dummies
  0-7645-5215-5

- Horses For Dummies
  0-7645-9797-3
- Jewelry Making & Beading For Dummies
  0-7645-2571-9
- Orchids For Dummies
  0-7645-6759-4
- Puppies For Dummies
  0-7645-5255-4
- Rock Guitar For Dummies
  0-7645-5356-9
- Sewing For Dummies
  0-7645-6847-7
- Singing For Dummies
  0-7645-2475-5

INTERNET & DIGITAL MEDIA

0-470-04529-9

0-470-04894-8

Also available:
- Blogging For Dummies
  0-471-77084-1
- Digital Photography For Dummies
  0-7645-9802-3
- Digital Photography All-in-One Desk Reference For Dummies
  0-470-03743-1
- Digital SLR Cameras and Photography For Dummies
  0-7645-9803-1
- eBay Business All-in-One Desk Reference For Dummies
  0-7645-8438-3
- HDTV For Dummies
  0-470-09673-X

- Home Entertainment PCs For Dummies
  0-470-05523-5
- MySpace For Dummies
  0-470-09529-6
- Search Engine Optimization For Dummies
  0-471-97998-8
- Skype For Dummies
  0-470-04891-3
- The Internet For Dummies
  0-7645-8996-2
- Wiring Your Digital Home For Dummies
  0-471-91830-X

SPORTS, FITNESS, PARENTING, RELIGION & SPIRITUALITY

0-471-76871-5

0-7645-7841-3

Also available:
- Catholicism For Dummies
  0-7645-5391-7
- Exercise Balls For Dummies
  0-7645-5623-1
- Fitness For Dummies
  0-7645-7851-0
- Football For Dummies
  0-7645-3936-1
- Judaism For Dummies
  0-7645-5299-6
- Potty Training For Dummies
  0-7645-5417-4
- Buddhism For Dummies
  0-7645-5359-3

- Pregnancy For Dummies
  0-7645-4483-7 †
- Ten Minute Tone-Ups For Dummies
  0-7645-7207-5
- NASCAR For Dummies
  0-7645-7681-X
- Religion For Dummies
  0-7645-5264-3
- Soccer For Dummies
  0-7645-5229-5
- Women in the Bible For Dummies
  0-7645-8475-8

TRAVEL

0-7645-7749-2

0-7645-6945-7

Also available:
- Alaska For Dummies
  0-7645-7746-8
- Cruise Vacations For Dummies
  0-7645-6941-4
- England For Dummies
  0-7645-4276-1
- Europe For Dummies
  0-7645-7529-5
- Germany For Dummies
  0-7645-7823-5
- Hawaii For Dummies
  0-7645-7402-7

- Italy For Dummies
  0-7645-7386-1
- Las Vegas For Dummies
  0-7645-7382-9
- London For Dummies
  0-7645-4277-X
- Paris For Dummies
  0-7645-7630-5
- RV Vacations For Dummies
  0-7645-4442-X
- Walt Disney World & Orlando
  For Dummies
  0-7645-9660-8

GRAPHICS, DESIGN & WEB DEVELOPMENT

0-7645-8815-X

0-7645-9571-7

Also available:
- 3D Game Animation For Dummies
  0-7645-8789-7
- AutoCAD 2006 For Dummies
  0-7645-8925-3
- Building a Web Site For Dummies
  0-7645-7144-3
- Creating Web Pages For Dummies
  0-470-08030-2
- Creating Web Pages All-in-One Desk
  Reference For Dummies
  0-7645-4345-8
- Dreamweaver 8 For Dummies
  0-7645-9649-7

- InDesign CS2 For Dummies
  0-7645-9572-5
- Macromedia Flash 8 For Dummies
  0-7645-9691-8
- Photoshop CS2 and Digital
  Photography For Dummies
  0-7645-9580-6
- Photoshop Elements 4 For Dummies
  0-471-77483-9
- Syndicating Web Sites with RSS Feed
  For Dummies
  0-7645-8848-6
- Yahoo! SiteBuilder For Dummies
  0-7645-9800-7

NETWORKING, SECURITY, PROGRAMMING & DATABASES

0-7645-7728-X

0-471-74940-0

Also available:
- Access 2007 For Dummies
  0-470-04612-0
- ASP.NET 2 For Dummies
  0-7645-7907-X
- C# 2005 For Dummies
  0-7645-9704-3
- Hacking For Dummies
  0-470-05235-X
- Hacking Wireless Networks
  For Dummies
  0-7645-9730-2
- Java For Dummies
  0-470-08716-1

- Microsoft SQL Server 2005 For Dumm
  0-7645-7755-7
- Networking All-in-One Desk Referen
  For Dummies
  0-7645-9939-9
- Preventing Identity Theft For Dummie
  0-7645-7336-5
- Telecom For Dummies
  0-471-77085-X
- Visual Studio 2005 All-in-One Desk
  Reference For Dummies
  0-7645-9775-2
- XML For Dummies
  0-7645-8845-1